Homology of
Analytic Sheaves
and Duality Theorems

CONTEMPORARY SOVIET MATHEMATICS
Series Editor: Revaz Gamkrelidze, *Steklov Institute, Moscow, USSR*

ASYMPTOTICS OF OPERATOR AND PSEUDO-DIFFERENTIAL EQUATIONS
V. P. Maslov and V. E. Nazaikinskii

COHOMOLOGY OF INFINITE-DIMENSIONAL LIE ALGEBRAS
D. B.Fuks

DIFFERENTIAL GEOMETRY AND TOPOLOGY
A. T. Fomenko

HOMOLOGY OF ANALYTIC SHEAVES AND DUALITY THEOREMS
V. D. Golovin

LINEAR DIFFERENTIAL EQUATIONS OF PRINCIPAL TYPE
Yu. V. Egorov

THE OBLIQUE DERIVATIVE PROBLEM OF POTENTIAL THEORY
A. I. Yanushauskas

OPTIMAL CONTROL
V. M. Alekseev, V. M. Tikhomirov, and S. V. Fomin

THEORY OF SOLITONS: The Inverse Scattering Method
S. Novikov, S. V. Manakov, L. P. Pitaevskii, and V. E. Zakharov

TOPICS IN MODERN MATHEMATICS: Petrovskii Seminar No. 5
Edited by O. A. Oleinik

Homology of Analytic Sheaves and Duality Theorems

V. D. Golovin
A. M. Gorky Kharkov State University
Kharkov, USSR

Translated from Russian by
Norman Stein

Consultants Bureau • New York and London

Library of Congress Cataloging in Publication Data

Golovin, V. D. (Viktor Dmitrievich)
    [Gomologii analiticheskikh puchkov i teoremy dvoĭstvennosti. English]
    Homology of analytic sheaves and duality theorems / V. D. Golovin; translated
from Russian by Norman Stein.
        p.   cm.—(Contemporary Soviet mathematics)
    Translation of: Gomologii analiticheskikh puchkov i teoremy dvoĭstvennosti.
    Bibliography: p.
    Includes index.
    ISBN-13: 978-1-4684-1679-4   e-ISBN-13: 978-1-4684-1677-0
    DOI: 10.1007/978-1-4684-1677-0
    1. Analytic sheaves. 2. Homology theory. 3. Duality theory (Mathematics) I. Title.
II. Series.
QA612.36.G6513   1989   89-458
514′.224—dc19   CIP

PREFACE

The homology of analytic sheaves is a natural apparatus in the theory of duality on complex spaces. The corresponding apparatus in algebraic geometry was developed by Grothendieck in the fifties. In complex analytic geometry the apparatus of homology was missing until recently, and in its stead the hypercohomology of complex sheaves (the hyper-Ext functors) and the Aleksandrov–Čech homology with coefficients in copresheaves were used. The homology of analytic sheaves, sheaves of germs of homology and homology groups of analytic sheaves, were introduced and studied in the mid-seventies in a number of papers by the author. The main goal of this book is to give a systematic and detailed account of the homology theory of analytic sheaves and some of its applications to duality theory on complex spaces and to the theory of hyperfunctions. In order to read this book one must be acquainted with the foundations of homological algebra and the theory of topological vector spaces. Only the most elementary concepts and results from the theory of functions of several complex variables are assumed to be known. The information needed about sheaves and complex spaces is recounted briefly at the beginning of the first chapter.

V. D. Golovin

CONTENTS

Chapter 1

ANALYTIC SHEAVES

1. Preliminary Information

1.1. By a *sheaf* of Abelian groups on a topological space X we mean a topological space $\mathscr{F}$, for which there is defined a surjective map $\pi \colon \mathscr{F} \to X$, satisfying the following conditions:

a) π is a local homeomorphism;

b) the stalk $\mathscr{F}_x = \pi^{-1}(x)$ is an Abelian group for each $x \in X$;

c) the group operations (defined stalkwise) are continuous in the topology of the space $\mathscr{F}$.[1]

1.1.1. By a *section* of the sheaf $\mathscr{F}$ over the open set $U \subset X$ we mean a continuous map $f \colon U \to \mathscr{F}$, for which $\pi \circ f = \mathrm{id}$ is the identity map of the set U onto itself.

The set $\Gamma(U; \mathscr{F})$ of all subsections of the sheaf $\mathscr{F}$ over the open set $U \subset X$ is an Abelian group, and the restriction map

$$r^U_V \colon \Gamma(U; \mathscr{F}) \to \Gamma(V; \mathscr{F})$$

is a homomorphism of Abelian groups, for any open sets $V \subset U$.

By the *germ* of the section $f \in \Gamma(U; \mathscr{F})$ at the point $x \in U$ is meant the image of f in the stalk $\mathscr{F}_x$ under the natural homomorphism $\Gamma(U; \mathscr{F}) \to \mathscr{F}_x$.

---

[1]See *Notes* on pages 165–198.

1.1.2. Example. Let X be a topological space, G be a discrete Abelian group, $\mathscr{F} = X \times G$ be the topological product, $\pi\colon \mathscr{F} \to X$ be the projection onto the first factor. Then $\mathscr{F}$ is a sheaf of Abelian groups on X, which is usually denoted by the letter G, and is called the *constant sheaf* with base X and stalk G. The group of sections $\Gamma(U; G)$ over the open set $U \subset X$ obviously consists of all locally constant maps $f\colon\ U \to G$.

1.1.3. Remark. One defines sheaves of rings and sheaves of modules over sheaves of rings analogously to sheaves of Abelian groups.

1.2. By a *presheaf* of Abelian groups on a topological space X is meant a contravariant functor from the category of open sets of the space X to the category of Abelian groups. In other words, a presheaf F on X associates with each open set $U \subset X$ an Abelian group $F(U)$, and with each pair of open sets $V \subset U$, a homomorphism of Abelian groups $r_V^U\colon F(U) \to F(V)$, such that $r_U^U = \mathrm{id}$ is the identity isomorphism of the group $F(U)$ onto itself, and $r_W^V \circ r_V^U = r_W^U$ for $W \subset V \subset U$.

1.2.1. A sheaf $\mathscr{F}$ on X *defines* a presheaf F, for which $F(U) = \Gamma(U; \mathscr{F})$, and r_V^U is the restriction homomorphism (cf. point 1.1.1). Conversely, with each presheaf F one can, in a natural way, associate a sheaf $\mathscr{F}$ for which $\mathscr{F}_x = \varinjlim F(U)$ for each $x \in X$, where U runs through all open neighborhoods of the point x (cf. Serre [3, p. 374]). One says that the sheaf $\mathscr{F}$ *is generated* by the presheaf F. However, it is not the case that each presheaf is defined by a sheaf.

A presheaf F on X is defined by a sheaf if and only if for each union $U = \cup\, U_i$ of open sets on X the following conditions hold:

a) if $f \in F(U)$ and $r_{U_i}^U(f) = 0$ for each i, then $f = 0$;

b) if $f_i \in F(U_i)$ are such that $r^{U_i}_{U_i \cap U_j}(f_i) = r^{U_j}_{U_i \cap U_j}(f_j)$ for each pair of indices (i, j), then there exists an $f \in F(U)$ for which $r^U_{U_i}(f) = f_i$ for each i (cf. Serre [3, pp. 374–375] and Godement [1, pp. 130–132]).

1.2.2. Examples. Let X be a topological space, G be an Abelian group, $F(U) = G$ for each open set $U \subset X$, and r^U_V be the identity automorphism of the group G for each pair of open sets $V \subset U$. Then F is a presheaf of Abelian groups on X. The sheaf generated by it is the constant sheaf of Example 1.1.2. On the other hand, if the space X is not connected, and the group G consists of at least two elements, then the presheaf F is not defined by any sheaf.

1.2.3. Let $F(U)$ be the set of all continuous (numerical) functions on the open set $U \subset X$, assuming complex values, and $r^U_V : F(U) \to F(V)$ for any pair of open sets $V \subset U$ be the restriction map. Then the presheaf F is defined by the sheaf $\mathscr{C}$, which is called the *sheaf of germs of continuous functions* on X.

1.2.4. One defines the *sheaf of germs of holomorphic functions* $\mathscr{O}$ (respectively the *sheaf of germs of C^∞ functions $\mathscr{E}$*) on the domain G of the space $\mathbf{C}^n$ analogously.

1.3. Let $\mathscr{F}$ be a sheaf of Abelian groups on a topological space X. An open set $\mathscr{F}' \subset \mathscr{F}$ is called a *subsheaf* of the sheaf $\mathscr{F}$ if $\pi(\mathscr{F}') = X$ and $\mathscr{F}'_x = \mathscr{F}' \cap \mathscr{F}_x$ for each $x \in X$ is a subgroup of the group $\mathscr{F}_x$. In this case the topological space $\mathscr{F}'$ itself is a sheaf of Abelian groups on X, for which the topology, projection, and group operations are induced from $\mathscr{F}$. Over each open set $U \subset X$ the group of sections $\Gamma(U; \mathscr{F}')$ is a subgroup of the group $\Gamma(U; \mathscr{F})$. Obviously this inclusion is compatible with the restriction homomorphisms.

The presheaf F'', which associates with each open set $U \subset X$ the quotient-group $F''(U) = \Gamma(U; \mathscr{F})/\Gamma(U; \mathscr{F}')$ and with each pair of open sets $V \subset U$ the homomorphism $r^U_V : F''(U) \to F''(V)$, induced by the restriction homomorphism $r^U_V: \Gamma(U; \mathscr{F}) \to \Gamma(V; \mathscr{F})$, generates a sheaf of Abelian groups $\mathscr{F}''$ on X. The sheaf $\mathscr{F}''$ is called the *quotient-sheaf* of the sheaf $\mathscr{F}$ by the subsheaf $\mathscr{F}'$ and is denoted by $\mathscr{F}/\mathscr{F}'$.

1.3.1. Let Y be a locally closed subspace of X. Then for any sheaf of Abelian groups $\mathscr{L}$ on Y there exists a unique sheaf of Abelian groups $\mathscr{L}^X$ on X, inducing $\mathscr{L}$ on Y and 0 on $X \setminus Y$ (cf. Godement [1, p. 161]).

1.3.2. Example. Let Y be a locally closed subspace of X. Then there exists a sheaf $\mathscr{F}_Y$ on X, which induces on Y the restriction $\mathscr{F} \mid Y$ and on $X \setminus Y$ the zero sheaf (cf. point 1.8 for the definition of the sheaf $\mathscr{F} \mid Y$). If U is an open set in X, then $\mathscr{F}_U$ is a subsheaf of the sheaf $\mathscr{F}$, and the sheaf $\mathscr{F}_{X \setminus U}$ is isomorphic to the quotient-sheaf $\mathscr{F}/\mathscr{F}_U$ (cf. Godement [1, pp. 160–162]).

1.4. Let $\mathcal{F}$ and $\mathcal{G}$ be sheaves of Abelian groups on the topological space X. A continuous map φ: $\mathcal{F} \to \mathcal{G}$ is called a *sheaf homomorphism*, if for each $x \in X$ it induces a homomorphism of Abelian groups φ_x: $\mathcal{F}_x \to \mathcal{G}_x$. Obviously such a map φ must be a local homeomorphism.

1.4.1. A homomorphism of sheaves φ: $\mathcal{F} \to \mathcal{G}$ defines for each open set $U \subset X$ a homomorphism of Abelian groups φ_U: $\Gamma(U; \mathcal{F}) \to \Gamma(U; \mathcal{G})$ such that $r^U_V \circ \varphi_U = \varphi_V \circ r^U_V$ for each pair of open sets $V \subset U$, i.e., the homomorphisms φ_U are compatible with the restriction homomorphisms. Conversely, each such collection of homomorphisms (φ_U) defines a homomorphism φ of sheaves of Abelian groups.

1.4.2. A homomorphism of sheaves of Abelian groups is called a *monomorphism* (respectively, an *epimorphism* or *isomorphism*), if it is injective (respectively, surjective or bijective).

1.4.3. Let φ: $\mathcal{F} \to \mathcal{G}$ be a homomorphism of sheaves of Abelian groups. Since the zero section defines an open set, the preimage $\varphi^{-1}(0)$ is a subsheaf of the sheaf $\mathcal{F}$; it is called the *kernel* of the homomorphism φ and is denoted by Ker φ.

Since φ is a local homeomorphism, $\varphi(\mathcal{F})$ is a subsheaf of the sheaf $\mathcal{G}$; it is called the *image* of the homomorphism φ and is denoted by Im φ.

1.4.4. A sequence of sheaves and homomorphisms of them

$$\mathcal{F}_1 \xrightarrow{\varphi_1} \mathcal{F}_2 \xrightarrow{\varphi_2} \cdots \xrightarrow{\varphi_n} \mathcal{F}_{n+1}$$

is said to be *exact*, if Im φ_i = Ker φ_{i+1} for each $i = 1, 2, ..., n - 1$. In particular, the sequence

$$0 \to \mathcal{F} \xrightarrow{\varphi} \mathcal{G} \xrightarrow{\psi} \mathcal{H} \dashrightarrow 0$$

is exact, if φ is a monomorphism and induces an isomorphism $\mathcal{F} = $ Im φ, and ψ is an epimorphism and induces an isomorphism Coker $\varphi = \mathcal{G} / $ Im $\varphi = \mathcal{H}$ (in other words, $\mathcal{F}$ can be considered as a subsheaf of the sheaf $\mathcal{G}$, and $\mathcal{H}$ as a quotient sheaf of $\mathcal{G}$ by the subsheaf $\mathcal{F}$).

1.5. Let $\mathcal{O}$ be a sheaf of rings with identity on the topological space X. Sheaves of left modules over the sheaf of rings $\mathcal{O}$ are called $\mathcal{O}$-*modules* for short. In particular, **Z**-modules are sheaves of Abelian groups.

Let $\mathscr{F}$ and $\mathscr{G}$ be two ϑ-modules. We denote by $\operatorname{Hom}_\vartheta(X; \mathscr{F}, \mathscr{G})$ the group of ϑ-homomorphisms $\mathscr{F} \to \mathscr{G}$. The sheaf $\mathscr{G}$ is said to be an *injective ϑ-module*, if the functor $\mathscr{F} \to \operatorname{Hom}_\vartheta(X; \mathscr{F}, \mathscr{G})$ is exact on the category of ϑ-modules. In other words, an ϑ-module $\mathscr{G}$ is injective, if the diagram with exact row

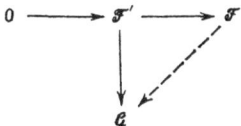

can be completed by the dashed arrow to a commutative diagram.

1.5.1. *Any ϑ-module is isomorphic to a submodule of an injective ϑ-module* (cf. Grothendieck [2, p. 63] and Godement [1, p. 291]).

1.5.2. By a *resolution* of an ϑ-module $\mathscr{F}$ is meant a right complex of ϑ-modules $\mathscr{L}^\cdot = (\mathscr{L}^k)_{k \geq 0}$, for which $H^0 \mathscr{L}^\cdot = \mathscr{F}$ and $H^k \mathscr{L}^\cdot = 0$ for $k \geq 1$. In other words, there is an exact sequence

$$0 \longrightarrow \mathscr{F} \longrightarrow \mathscr{L}^0 \longrightarrow \mathscr{L}^1 \longrightarrow \cdots$$

of ϑ-modules and homomorphisms of them. The resolution $\mathscr{L}^\cdot$ is said to be *injective* if all the ϑ-modules $\mathscr{L}^k$ ($k = 0, 1, \ldots$) are injective.

It follows from point 1.5.1 that for any ϑ-module there exists an injective resolution.

1.6. Let T be a left exact covariant functor from the category of ϑ-modules to the category of Abelian groups. Then there are defined the right *derived functors* of the functor T:

$$R^k T (\mathscr{F}) = H^k T (\mathscr{L}^\cdot),$$

where $\mathscr{L}^\cdot$ is an injective resolution of the ϑ-module $\mathscr{F}$, while this definition is independent of the choice of injective resolution $\mathscr{L}^\cdot$ up to a natural isomorphism (cf. Cartan and Eilenberg [1, p. 110]).

The derived functors are covariant functors from the category of ϑ-modules to the category of Abelian groups and have the following properties:

a) the functor $R^0 T$ is isomorphic to the functor T;

b) for each exact sequence of $\mathcal{O}$-modules

$$0 \longrightarrow \mathcal{F}' \longrightarrow \mathcal{F} \longrightarrow \mathcal{F}'' \longrightarrow 0$$

there is defined a connecting homomorphism

$$\delta:\ R^kT\,(\mathcal{F}'') \longrightarrow R^{k+1}T\,(\mathcal{F}')$$

such that the sequence of Abelian groups

$$0 \longrightarrow T\,(\mathcal{F}') \longrightarrow T\,(\mathcal{F}) \longrightarrow T\,(\mathcal{F}'') \overset{\delta}{\longrightarrow} R^1T\,(\mathcal{F}') \longrightarrow \cdots$$
$$\cdots \longrightarrow R^kT\,(\mathcal{F}') \longrightarrow R^kT\,(\mathcal{F}) \longrightarrow R^kT\,(\mathcal{F}'') \overset{\delta}{\longrightarrow} R^{k+1}T\,(\mathcal{F}') \longrightarrow \cdots$$

is exact;

c) if there is a commutative diagram with exact rows

$$\begin{array}{ccccccccc} 0 & \longrightarrow & \mathcal{F}' & \longrightarrow & \mathcal{F} & \longrightarrow & \mathcal{F}'' & \longrightarrow & 0 \\ & & \downarrow & & \downarrow & & \downarrow & & \\ 0 & \longrightarrow & \mathcal{G}' & \longrightarrow & \mathcal{G} & \longrightarrow & \mathcal{G}'' & \longrightarrow & 0 \end{array}$$

then the diagram

$$\begin{array}{ccc} R^kT\,(\mathcal{F}'') & \overset{\delta}{\longrightarrow} & R^{k+1}T\,(\mathcal{F}') \\ \downarrow & & \downarrow \\ R^kT\,(\mathcal{G}'') & \overset{\delta}{\longrightarrow} & R^{k+1}T\,(\mathcal{G}') \end{array}$$

is commutative;

d) $R^kT(\mathcal{L}) = 0$ for $k \geq 1$, if the $\mathcal{O}$-module $\mathcal{L}$ is injective.

1.7. Let $\mathcal{F}$ be a sheaf of Abelian groups on the topological space X. By the *support* of a section $f \in \Gamma(X;\ \mathcal{F})$ is meant the closed set Supp f, consisting of all points $x \in X$, for which $f_x \neq 0$.

By a *family of supports* in a topological space X is meant a set Φ of closed subsets of X, satisfying the following conditions:

a) if $S_1, S_2 \in \Phi$, then $S_1 \cup S_2 \in \Phi$;

b) if $S_1 \in \Phi$ and S_2 is a closed subset of S_1, then $S_2 \in \Phi$.

The set $\Gamma_\Phi(X; \mathscr{F})$ of all sections $f \in \Gamma(X; \mathscr{F})$, for which $\mathrm{Supp}\, f \in \Phi$, is a subgroup of the group $\Gamma(X; \mathscr{F})$. In this way there is defined a co-variant left exact functor $\mathscr{F} \mapsto \Gamma_\Phi(X; \mathscr{F})$ from the category of sheaves of Abelian groups on X to the category of Abelian groups. Its right derived functors

$$H_\Phi^k(X; \mathscr{F}) = R^k\Gamma_\Phi(X; \mathscr{F})$$

are called the *cohomology groups* of the space X with coefficients in the sheaf $\mathscr{F}$ and with supports from the family Φ (or, for short, the cohomology groups of the sheaf $\mathscr{F}$ with supports in Φ).[2]

1.7.1. Let $\mathscr{F}$ be a sheaf of Abelian groups on the topological space X, Φ be a family of supports in X, $\mathscr{L}^\cdot$ be a Φ-acyclic resolution of the sheaf $\mathscr{F}$, i.e.,

$$H_\Phi^p(X; \mathscr{L}^q) = 0 \quad \text{for} \quad p \geqslant 1, \quad q \geqslant 0.$$

Then there exist natural isomorphisms

$$H_\Phi^k(X; \mathscr{F}) = H^k\Gamma_\Phi(X; \mathscr{L}^\cdot)$$

("formal de Rham theorem"; cf. Godement [1, p. 206]).

1.7.2. Examples. A sheaf of Abelian groups $\mathscr{L}$ over X is said to be *flabby*, if for any open set $U \subset X$ the restriction homomorphism $r^X{}_U$: $\Gamma(X; \mathscr{L}) \to \Gamma(U; \mathscr{L})$ is surjective, i.e., if any section of the sheaf $\mathscr{L}$ over an arbitrary open set can be extended to all of X. Any injective sheaf of Abelian groups (i.e., of Z-modules) is a flabby sheaf. For a flabby sheaf $\mathscr{L}$ and any family of supports Φ in X,

$$H_\Phi^k(X; \mathscr{L}) = 0 \quad \text{for} \quad k \geqslant 1$$

(cf. Godement [1, p. 171]).

1.7.3. Let $\mathscr{F}$ be a sheaf of Abelian groups on X, $\mathscr{C}^0(\mathscr{F})$ be the sheaf of germs of all, not necessarily continuous, "sections" of the sheaf $\mathscr{F}$ and $\mathscr{X}^0 = \mathscr{F}$. By induction we set $\mathscr{C}^k(\mathscr{F}) = \mathscr{C}^0(\mathscr{X}^k)$, $\mathscr{X}^{k+1} = \mathscr{C}^k(\mathscr{F})/\mathscr{X}^k$. We get an exact sequence, consisting of flabby sheaves,

$$0 \to \mathscr{F} \to \mathscr{C}^0(\mathscr{F}) \to \mathscr{C}^1(\mathscr{F}) \to \ldots,$$

which is called the *canonical resolution* of the sheaf $\mathcal{F}$ (cf. Godement [1, p. 193]). By definition there are natural isomorphisms

$$H^k_\Phi(X;\ \mathcal{F}) = H^k \Gamma_\Phi(X;\ \mathcal{C}^\bullet(\mathcal{F})).$$

1.7.4. A sheaf of Abelian groups $\mathcal{L}$ on a paracompact topological space X is said to be *soft*, if any section of the sheaf $\mathcal{L}$ over an arbitrary closed set can be extended to all of X. Any flabby sheaf on X is soft. For a soft sheaf $\mathcal{L}$

$$H^k(X;\ \mathcal{L}) = 0 \quad \text{for} \quad k \geqslant 1$$

(cf. Godement [1, p. 177]). Thus, one can calculate the cohomology groups of sheaves with the help of soft resolutions. In view of Poincaré's lemma, de Rham's theorem follows directly from this (cf. de Rham [1, p. 145] and Godement [1, p. 207]).

1.7.5. A sheaf of Abelian groups $\mathcal{L}$ on a paracompact topological space X is said to be *fine*, if for any closed sets $A, B \subset X$, $A \cap B = \varnothing$, there exists a homomorphism $\mathcal{L} \to \mathcal{L}$, which induces the identity map in a neighborhood of the set A and 0 in a neighborhood of the set B. Any fine sheaf is soft. For any sheaf of Abelian groups $\mathcal{F}$ on X, the sheaf $\mathcal{C}^0(\mathcal{F})$ is fine. If $\mathcal{L}$ is a fine sheaf, then for any locally finite open covering $\mathfrak{u} = (U_i)$ of the space X there exists a family of homomorphisms $\xi_i \colon \mathcal{L} \to \mathcal{L}$, satisfying the following conditions:

a) for each i, in some neighborhood of the set $X \setminus U_i$ the homomorphism ξ_i induces the zero map (i.e., Supp $\xi_i \subset U_i$);

b) $\sum_i \xi_i = \mathrm{id}$ is the identity automorphism of the sheaf $\mathcal{L}$.[3]

1.8. Let $\mathcal{F}$ be a sheaf of Abelian groups on a topological space X. Then for any arbitrary subspace $Y \subset X$ the topological space

$$\mathcal{F}\,|\,Y = \pi^{-1}(Y)$$

is a sheaf of Abelian groups on Y; it is called *the restriction* of the sheaf $\mathcal{F}$ to Y (or the induced sheaf).

For an arbitrary family of supports Φ in Y one usually sets, for short,

$$H^k_\Phi(Y;\ \mathcal{F}) = H^k_\Phi(Y;\ \mathcal{F}\,|\,Y).$$

1.8.1. Let U be an open set in X and $S = X \setminus U$ be the complementary closed set. Then for the groups of sections with compact supports of an arbitrary flabby sheaf $\mathscr{L}$ over X there is an exact sequence

$$0 \to \Gamma_c(U;\ \mathscr{L}) \to \Gamma_c(X;\ \mathscr{L}) \to \Gamma_c(S;\ \mathscr{L}) \to 0.$$

From this one gets the cohomology exact sequence connected with the closed subset S:

$$\ldots \to H_c^k(U;\mathscr{F}) \to H_c^k(X;\mathscr{F}) \to H_c^k(S;\mathscr{F}) \to H_c^{k+1}(U;\mathscr{F}) \to \ldots$$

(cf. Godement [1, p. 216]).

1.8.2. Analogously there is an exact sequence of groups of sections

$$0 \to \Gamma_S(X;\ \mathscr{L}) \to \Gamma(X;\ \mathscr{L}) \to \Gamma(U;\ \mathscr{L}) \to 0,$$

from which one gets an exact cohomology sequence connected with an open subset U:

$$\ldots \to H_S^k(X;\mathscr{F}) \to H^k(X;\mathscr{F}) \to H^k(U;\mathscr{F}) \to H_S^{k+1}(X;\mathscr{F}) \to \ldots$$

The groups $H_S^k(X;\mathscr{F})$ are called the *local cohomology* groups with respect to the set S. The presheaf $U \mapsto H^k_{S \cap U}(U;\ \mathscr{F})$ generates a sheaf of Abelian groups $\mathscr{H}_S^k(\mathscr{F})$ over X, which is called the *sheaf of germs of local cohomology* of the sheaf $\mathscr{F}$ with respect to the closed set S. There is a spectral sequence

$$E_2^{p,\ q} = H^p(X;\ \mathscr{H}_S^q(\mathscr{F})) \Rightarrow H_S^{p+q}(X;\ \mathscr{F})$$

(cf. Grothendieck [4]).

1.9. Let $\mathscr{F}$ be a sheaf of Abelian groups on a topological space X. For an arbitrary open covering $\mathfrak{U} = (U_i)_{i \in I}$ of the space X we set

$$C^k(\mathfrak{U};\ \mathscr{F}) = \prod \Gamma(U_{i_0 \ldots i_k};\ \mathscr{F}),$$

where the product is taken over those collections of indices $(i_0, \ldots, i_k)$, for which the intersection $U_{i_0 \ldots i_k} = U_{i_0} \cap \ldots \cap U_{i_k}$ is nonempty. The elements $f = (f_{i_0 \ldots i_k})$ of the group $C^k(\mathfrak{U};\ \mathscr{F})$ are called *cochains* of degree k. We define a homomorphism

$$\delta:\ C^k(\mathfrak{U};\ \mathscr{F}) \to C^{k+1}(\mathfrak{U};\ \mathscr{F}),$$

which acts according to the formula

$$(\delta f)_{i_0 \,\cdots\, i_{k+1}} = \sum_{s=0}^{k+1} (-1)^s r_{U_{i_0 \,\cdots\, i_{k+1}}}^{U_{i_0 \,\cdots\, \hat{i}_s \,\cdots\, i_{k+1}}} \left(f_{i_0 \,\cdots\, \hat{i}_s \,\cdots\, i_{k+1}} \right)$$

and is called the *coboundary operator* ($\hat{i_s}$ means that it is necessary to omit the index i_s). Since $\delta \circ \delta = 0$, we get a cochain complex of Abelian groups $C^{\cdot}(\mathfrak{U}; \mathscr{F})$, whose cohomology groups

$$H^k(\mathfrak{U}; \mathscr{F}) = H^k C^{\cdot}(\mathfrak{U}; \mathscr{F})$$

are called the *cohomology groups of the covering* $\mathfrak{U}$ with coefficients in the sheaf $\mathscr{F}$.

Remark. Instead of a sheaf one can consider a presheaf.

1.9.1. Let $\mathfrak{V} = (V_j)_{j \in J}$ be an open covering of the space X, which refines the covering $\mathfrak{U}$. Then one can define a map $\theta \colon J \to I$ such that $V_j \subset U_{\theta(j)}$ for each $j \in J$. Setting

$$(\theta f)_{i_0 \,\cdots\, i_k} = r_{V_{j_0 \,\cdots\, j_k}}^{U_{\theta(j_0) \,\cdots\, \theta(i_k)}} \left(f_{\theta(j_0) \,\cdots\, \theta(i_k)} \right),$$

we get homomorphisms of cochain groups

$$\theta \colon C^k(\mathfrak{U}; \mathscr{F}) \to C^k(\mathfrak{V}; \mathscr{F}),$$

compatible with the coboundary operators, i.e., a cochain map (or homomorphism of cochain complexes). Two such homomorphisms, corresponding to different choices of maps $\theta \colon J \to I$, are cochain homotopic. Thus we get a homomorphism of cohomology groups

$$\theta^{\mathfrak{U}}_{\mathfrak{V}} \colon H^k(\mathfrak{U}; \mathscr{F}) \to H^k(\mathfrak{V}; \mathscr{F}),$$

which is independent of the choice of the map $\theta \colon J \to I$.

1.9.2. Two coverings are said to be *equivalent* if they refine one another. A covering is said to be *proper* if its set of indices is a subset of the set $\mathfrak{P}(X)$ of all parts of the space X. Each covering is equivalent to a proper covering. The proper coverings of the space X form a set which is ordered and filtered to the left by the relation of being a refinement.

The homomorphisms $\theta^{\mathfrak{U}}_{\mathfrak{V}}$ satisfy the following conditions:

a) $\theta^{\mathfrak{U}}_{\mathfrak{U}} = \mathrm{id}$ is the identity isomorphism;

b) $\theta^{\mathfrak{B}}_{\mathfrak{W}} \circ \theta^{\mathfrak{U}}_{\mathfrak{B}} = \theta^{\mathfrak{U}}_{\mathfrak{W}}$, if $\mathfrak{B}$ is a refinement of $\mathfrak{U}$, and $\mathfrak{W}$ is a refinement of $\mathfrak{B}$. In particular, for equivalent coverings $\mathfrak{U}$ and $\mathfrak{B}$ the homomorphisms $\theta^{\mathfrak{U}}_{\mathfrak{B}}$ and $\theta^{\mathfrak{B}}_{\mathfrak{U}}$ are mutually inverse isomorphisms.

Thus, the groups $H^k(\mathfrak{U}; \mathcal{F})$ and homomorphisms $\theta^{\mathfrak{U}}_{\mathfrak{B}}$ form an inductive system. Consequently, one can define the inductive limit over the filtered set of all proper open coverings

$$\check{H}^k(X; \mathcal{F}) = \varinjlim H^k(\mathfrak{U}; \mathcal{F}).$$

This limit is called the *Aleksandrov–Čech cohomology group* of the space X with coefficients in the sheaf $\mathcal{F}$.[4]

1.9.3. Let $\mathfrak{U}$ be an arbitrary open covering of the space X. Then there is the *Leray spectral sequence* for the covering

$$E_2^{p,\,q} = H^p(\mathfrak{U}; \mathcal{H}^q(\mathcal{F})) \Rightarrow H^{p+q}(X; \mathcal{F}),$$

where $\mathcal{H}^q(\mathcal{F})$ is the presheaf $U \mapsto H^q(U; \mathcal{F})$.

In particular, if the covering $\mathfrak{U}$ is acyclic, i.e., if for any $i_0, \ldots, i_p$

$$H^q(U_{i_0 \ldots i_p}; \mathcal{F}) = 0 \quad \text{for} \quad q \geqslant 1,$$

then the natural homomorphisms

$$H^k(\mathfrak{U}; \mathcal{F}) \to H^k(X; \mathcal{F})$$

are isomorphisms (cf. Godement [1, p. 240]).

1.9.4. *If the space X is paracompact, then the natural homomorphisms*

$$\check{H}^k(X; \mathcal{F}) \to H^k(X; \mathcal{F})$$

are isomorphisms (cf. Godement [1, pp. 256–257]).

1.10. Let $\mathcal{O}$ be a sheaf of rings with identity on the topological space X. For an arbitrary family of supports Φ in X we denote by $\mathrm{Hom}_{\mathcal{O},\Phi}(X; \mathcal{F}, \mathcal{G})$ the group of homomorphisms of $\mathcal{O}$-modules $\mathcal{F} \to \mathcal{G}$ with supports in Φ. We get a covariant left exact functor $\mathcal{G} \mapsto \mathrm{Hom}_{\mathcal{O},\Phi}(X; \mathcal{F}, \mathcal{G})$ from the category of $\mathcal{O}$-modules on X to the category of Abelian groups. We consider its right derived functors[5]

$$\operatorname{Ext}^{k}_{6,\,\Phi}(X;\ \mathscr{F},\ \mathscr{G}) = R^{k}\operatorname{Hom}_{6,\,\Phi}(X;\ \mathscr{F},\ \mathscr{G})$$

1.10.1. Like any derived functor $\mathscr{G} \mapsto \operatorname{Ext}^{k}_{6,\Phi}(X;\ \mathscr{F},\mathscr{G})$ is a (covariant) cohomology functor in the sense of point 1.6. Since for an injective 6-module $\mathscr{L}$ the functor $\mathscr{F} \to \operatorname{Hom}_{6,\Phi}(X;\ \mathscr{F},\mathscr{L})$ is exact, the functor $\mathscr{F} \to \operatorname{Ext}^{k}_{6,\Phi}(X;\ \mathscr{F},\mathscr{G})$ is a (contravariant) cohomology functor.

1.10.2. For an arbitrary open set $U \subset X$ we set, for brevity,

$$\operatorname{Ext}^{k}_{6}(U;\ \mathscr{F},\ \mathscr{G}) = \operatorname{Ext}^{k}_{6\,|\,U}(U;\ \mathscr{F}\,|\,U,\ \mathscr{G}\,|\,U).$$

The presheaf $U \mapsto \operatorname{Ext}^{k}_{6}(U;\ \mathscr{F},\mathscr{G})$ generates a sheaf $\mathscr{E}\mathrm{xt}_{6}^{k}(\mathscr{F},\mathscr{G})$. For $k = 0$ we get, in particular, the sheaf of germs of homomorphisms $\mathscr{H}\mathrm{om}_{6}(\mathscr{F},\mathscr{G})$.

There exists a spectral sequence

$$E_{2}^{p,\,q} = H^{p}_{\Phi}(X;\ \mathscr{E}\mathrm{xt}^{q}_{6}(\mathscr{F},\ \mathscr{G})) \Rightarrow \operatorname{Ext}^{p+q}_{6,\,\Phi}(X;\ \mathscr{F},\ \mathscr{G})$$

(cf. Godement [1, p. 295]).

1.11. The topological space X is called a *ringed space* if there is given on X a sheaf 6_X of commutative complex algebras with identity. The sheaf 6_X is called the *structure sheaf* of the ringed space X.

A continuous map $\varphi: X \to Y$ of ringed spaces is called a *morphism* if there is given a homomorphism of sheaves of complex algebras $\psi: \varphi^* 6_Y \to 6_X$. Here we denoted by $\varphi^* 6_Y$ the topological space

$$X \times_Y 6_Y = \{(x,\ f) \in X \times 6_Y\colon \varphi(x) = \pi(f)\},$$

which is a sheaf of complex algebras on X and is called the *inverse image* of the sheaf 6_Y. A morphism of ringed spaces is called an *isomorphism*, if φ is a homeomorphism and ψ is an isomorphism of sheaves of complex algebras.

Each open subset U in a ringed space X can be considered as a ringed space with structure sheaf $6_U = 6_X\,|\,U$.

1.11.1. Let G be a domain in the space $\mathbf{C}^n$ and 6_G be the sheaf of germs of holomorphic functions on G. We get an example of a ringed space which we shall also call a *domain* in $\mathbf{C}^n$.

We consider a more general example. Let M be a subset of G, defined by equations

$$f_i(z_1, \ldots, z_n) = 0 \qquad (1 \leqslant i \leqslant k),$$

where $f_1, \ldots, f_k$ are holomorphic functions on G. We set

$$\mathcal{O}_M = \mathcal{O}_G/\mathcal{I} \mid M,$$

where $\mathcal{I}$ is the subsheaf of ideals in $\mathcal{O}_G$ generated by the functions $f_i, \ldots, f_k$. We get a ringed space which we shall call an *analytic set* in G.

An analytic set is called *reduced* if $\mathcal{I}$ is the sheaf of germs of all holomorphic functions equal to zero on M.

1.11.2. By a *complex space* is meant a separated topological space X, endowed with a structure sheaf of complex algebras $\mathcal{O}_X$ (i.e., which is a ringed space) such that each point in X has an open neighborhood, isomorphic as a ringed space to an analytic set in a domain of the space $\mathbb{C}^n$.[6]

Thus, in relation to a complex space, analytic sets play the role of local models. If as local models one takes domains in the space $\mathbb{C}^n$ (respectively reduced analytic sets), then we get the definition of a *complex manifold* (respectively *reduced complex space*). A complex space is reduced if and only if the stalks of the structure sheaf do not contain nilpotent elements.

By a *morphism* of complex spaces is meant a morphism of the corresponding ringed spaces. A morphism of reduced complex spaces is called a *holomorphic map*.

1.11.3. By an *analytic sheaf* on a complex space X is meant an arbitrary sheaf of $\mathcal{O}_X$-modules on X.

An analytic sheaf $\mathcal{F}$ on X is said to be *coherent* if for each point $x \in X$ there exists an open neighborhood U, over which there is an exact sequence of $\mathcal{O}_U$-modules:

$$\mathcal{O}_U^r \to \mathcal{O}_U^p \to \mathcal{F} \mid U \to 0.$$

If in the exact sequence of analytic sheaves

$$0 \to \mathcal{F} \to \mathcal{G} \to \mathcal{H} \to 0$$

two of the three sheaves $\mathscr{F}, \mathscr{G}, \mathscr{H}$ are coherent, then the third sheaf is also coherent (cf. Serre [3, p. 382]).

Examples. Let M be an analytic set in the domain G of the space $\mathbf{C}^n$ (in the sense of point 1.11.1). Then the sheaf $\mathscr{F}$ of all germs of holomorphic functions equal to zero on M is a coherent subsheaf of ideals in $\mathcal{O}_G$ (*Cartan's theorem*; cf. Gunning and Rossi [1, p. 175]).

On the other hand, an arbitrary subsheaf of ideals $\mathscr{F}$ in $\mathcal{O}_G$ is coherent if and only if it is locally generated by a finite number of its sections (*Oka's theorem*; cf. Gunning and Rossi [1, p. 170]).[7]

1.11.4. Let X be a complex space. Since the stalk of the structure sheaf $\mathcal{O}_{X,x}$ is a local ring, it makes sense to speak of the value $f(x)$, assumed by the section $f \in \Gamma(X; \mathcal{O}_X)$ at the point $x \in X$. For this reason, sections of the sheaf $\mathcal{O}_X$ are often called *holomorphic functions*.

A subset $M \subset X$ is called an *analytic set* if in a neighborhood of each point $x \in X$ it can be defined as the set of common zeros of a finite number of holomorphic functions. To each analytic set corresponds a coherent subsheaf of ideals $\mathscr{F}$ in $\mathcal{O}_X$, consisting of the germs of all holomorphic functions equal to zero on M.

Conversely, each coherent subsheaf of ideals $\mathscr{F}$ in $\mathcal{O}_X$ defines an analytic set

$$M = \{x \in X: \mathfrak{I}_x \neq \mathcal{O}_{X,x}\},$$

which is called *the set of zeros* of the sheaf $\mathscr{F}$.

1.12. The complex space X is said to be *holomorphically convex* if for each compact set $K \subset X$ the set

$$\hat{K} = \{x \in X: |f(x)| \leqslant \sup_{y \in K} |f(y)| \quad \text{for all} \quad f \in \Gamma(X; \mathcal{O}_X)\}$$

is also compact. The space X is called *holomorphically separable* if for any two points $x_1 \neq x_2$ of X there exists a holomorphic function $f \in \Gamma(X; \mathcal{O}_X)$, for which $f(x_1) \neq f(x_2)$. A complex space is said to be *holomorphically complete* (or a *Stein space*), if it is holomorphically convex and holomorphically separated.[8]

1.12.1. Examples. The space $\mathbf{C}^n$ is holomorphically complete.

1.12.2. Any analytic set in the space C^n is a holomorphically complete space.

1.12.3. A domain G in the space C^n is a holomorphically complete space if and only if G is a domain of holomorphy (*Cartan–Thullen theorem*; cf. Vladimirov [1, p. 168]).

1.13. Let $\mathscr{F}$ be a coherent analytic sheaf on the holomorphically complete complex space X. Then the following two assertions hold (*Cartan's fundamental theorems (A) and (B)*; cf. Cartan [4]):

(A) $\Gamma(X; \mathscr{F})$ for each $x \in X$ generates the stalk $\mathscr{F}_x$ as an $\mathscr{O}_{X,x}$-module;

(B) $H^k(X; \mathscr{F}) = 0$ for $k \geq 1$.

(cf. Gunning and Rossi [1, p. 310]).[9]

1.14. Let $\varphi: X \to Y$ be a continuous map of topological spaces. Then for each sheaf of Abelian groups $\mathscr{F}$ on X there is defined a presheaf $U \mapsto \Gamma(\varphi^{-1}(U); \mathscr{F})$ on Y. This presheaf is defined by a sheaf of Abelian groups $\varphi_* \mathscr{F}$ on Y, i.e., $\Gamma(\varphi^{-1}(U); \mathscr{F}) = \Gamma(U; \varphi_* \mathscr{F})$ for any open set $U \subset Y$. The sheaf $\varphi_* \mathscr{F}$ is called the *direct image* of the sheaf $\mathscr{F}$ with respect to the map φ.

1.14.1. The correspondence $\varphi_*: \mathscr{F} \mapsto \varphi_* \mathscr{F}$ is a left exact covariant functor from the category of sheaves of Abelian groups on X to the category of sheaves of Abelian groups on Y. One defines the right derived functors $R^k \varphi_*$ ($k \geq 0$) of the functor φ_* in the usual way. For each sheaf of Abelian groups $\mathscr{F}$ on X the sheaf of Abelian groups $R^k \varphi_* \mathscr{F}$ on Y is generated by the presheaf $U \mapsto H^k(\varphi^{-1}(U); \mathscr{F})$ and for $k \geq 1$ it is called a *higher direct image* of the sheaf $\mathscr{F}$.

1.14.2. For an arbitrary continuous map of topological spaces $\varphi: X \to Y$ and any sheaf of Abelian groups $\mathscr{F}$ on X there is the *Leray spectral sequence* of the continuous map:

$$E_2^{p,q} = H^p(Y; R^q \varphi_* \mathscr{F}) \Rightarrow H^{p+q}(X; \mathscr{F})$$

(cf. Grothendieck [2, pp. 93–94]).

1.15. The continuous map φ: $X \to Y$ of locally compact topological spaces is called *proper* if for each compact set $K \subset Y$ its preimage $\varphi^{-1}(K)$ is a compact set in X (cf. Bourbaki [1, p. 121]).

1.15.1. There is the following *Grauert theorem on direct images*:

Let φ: $X \to Y$ be a proper morphism of complex spaces. Then for each coherent analytic sheaf $\mathscr{F}$ on X, its direct images $R^k \varphi_* \mathscr{F}$ $(k \geq 0)$ *are* coherent analytic sheaves on Y (cf. Grauert [4]).[10]

1.15.2. In the special case when the space Y consists of one point, one gets the *Cartan–Serre finiteness theorem*:

If the complex space X is compact and $\mathscr{F}$ is a coherent analytic sheaf on X, then the cohomology vector spaces $H^k(X; \mathscr{F})$ $(k \geq 0)$ are finite-dimensional (cf. Cartan and Serre [1]).

2. Injectivity Test[11]

2.1. Let $\mathscr{O}$ be a sheaf of commutative rings with units over the topological space X. One says that the $\mathscr{O}$-module $\mathscr{G}$ is an *essential extension* of the $\mathscr{O}$-module $\mathscr{F}$, if $\mathscr{F}$ is an $\mathscr{O}$-submodule of the module $\mathscr{G}$ and for any $\mathscr{O}$-submodule $\mathscr{S} \subset \mathscr{G}$ the condition $\mathscr{S} \cap \mathscr{F} = 0$ implies $\mathscr{S} = 0$.

Example. Let X be a complex space, $\mathscr{O}_X$ be its structure sheaf of complex algebras, $\mathscr{M}_X$ be *the sheaf of germs of meromorphic functions* on X, i.e., the sheaf of complete rings of fractions of the sheaf of rings $\mathscr{O}_X$. Then $\mathscr{O}_X \subset \mathscr{M}_X$ is an essential extension of $\mathscr{O}_X$-modules.

Lemma. Let $\mathscr{O}$ be a sheaf of commutative rings with unit over the topological space X. Then an extension of $\mathscr{O}$-modules $\mathscr{F} \subset \mathscr{G}$ is essential if and only if for each open set $U \subset X$ and each section $g \in \Gamma(U; \mathscr{G})$, different from zero, there exists a point $x \in U$ and a germ $\varphi_x \in \mathscr{O}_x$, such that $\varphi_x g_x \neq 0$ and $\varphi_x g_x \in \mathscr{F}_x$.

Proof. Obviously the essentialness of an extension is a local property. In other words, an extension of $\mathscr{O}$-modules $\mathscr{F} \subset \mathscr{G}$ is essential if and only if, for each sufficiently small open set $U \subset X$ the restriction $\mathscr{G}|U$ is an essential extension for $\mathscr{F} \mid U$ over the sheaf of rings $\mathscr{O} \mid U$. Let $\mathscr{F} \subset \mathscr{G}$ be an essential extension of $\mathscr{O}$-modules. Then for each section $g \in \Gamma(U; \mathscr{G})$, different from zero, we get an $\mathcal{O}\,|\,U$-submodule $(\mathcal{O}\,|\,U)g \neq 0$ in $\mathcal{G}\,|\,U$. Consequently, $(\mathcal{O}\,|\,U)g \cap (\mathcal{F}\,|\,U) \neq 0$, i.e., there exists a point $x \in U$, for which $\mathcal{O}_x g_x \cap \mathcal{F}_x \neq 0$. Conversely, let the extension of $\mathcal{O}$-modules $\mathcal{F} \subset \mathcal{G}$ not be essential. Then there exists a nonzero submodule $\mathcal{S} \subset \mathcal{G}$, for which $\mathcal{S} \cap \mathcal{F} = 0$. Over some open set $U \subset X$ there exists a section $s \in \Gamma(U; \mathcal{S})$, which is different from zero. By the assumption made, $(\mathcal{O}\,|\,U)s \cap (\mathcal{F}|U) = 0$, i.e., for each point $x \in U$ and each germ $\varphi_x \in \mathcal{O}_x$ either $\varphi_x s_x = 0$ or $\varphi_x s_x \notin \mathcal{F}_x$. The lemma is proved.

2.2. Proposition. Let $\mathcal{O}$ be a sheaf of commutative rings with unit on the topological space X. Then in order that the $\mathcal{O}$-module $\mathcal{F}$ be injective, it is necessary and sufficient that it not have proper essential extensions.

Proof. Let the $\mathcal{O}$-module $\mathcal{F}$ be injective and let $\mathcal{F} \subset \mathcal{G}$ be an essential extension of it. By the definition of injectivity the diagram with exact row

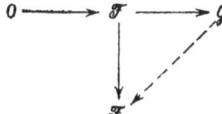

can be completed by the dashed arrow to a commutative diagram. Thus, the identity isomorphism of the $\mathcal{O}$-module $\mathcal{F}$ onto itself splits into the composition of a monomorphism $\mathcal{F} \to \mathcal{G}$ (inclusion) and a homomorphism $\mathcal{G} \to \mathcal{F}$ (the dashed arrow). The kernel of the latter will be denoted by $\mathcal{K}$. Then obviously $\mathcal{K} \cap \mathcal{F} = 0$, so $\mathcal{K} = 0$. This means that $\mathcal{F} = \mathcal{G}$. Conversely, let the $\mathcal{O}$-module $\mathcal{F}$ have no proper essential extensions. Let $\mathcal{F} \subset \mathcal{G}$ be an extension of the $\mathcal{O}$-module $\mathcal{F}$. We consider the set $\mathfrak{S}$ of all $\mathcal{O}$-submodules $\mathcal{S} \subset \mathcal{G}$ for which $\mathcal{S} \cap \mathcal{F} = 0$. We order the set $\mathfrak{S}$ by inclusion. Obviously the ordered set $\mathfrak{S}$ is inductive. By Zorn's lemma it contains a maximal element $\mathcal{M}$. Then the composition of the homomorphisms $\mathcal{F} \to \mathcal{G}$ and $\mathcal{G} \to \mathcal{G}/\mathcal{M}$ is a monomorphism, and $\mathcal{G}/\mathcal{M}$ can be considered as an essential extension of the $\mathcal{O}$-module $\mathcal{F}$. On the other hand, $\mathcal{F}$ has no proper essential extensions, so the homomorphism $\mathcal{F} \to \mathcal{G}/\mathcal{M}$ is an isomorphism of $\mathcal{O}$-modules. Thus, $\mathcal{F} + \mathcal{M} = \mathcal{G}$ and $\mathcal{F} \cap \mathcal{M} = 0$, i.e., $\mathcal{F}$ is a direct summand in $\mathcal{G}$. Since the $\mathcal{O}$-module $\mathcal{F}$ is contained in an injective $\mathcal{O}$-module, it is itself injective. The proposition is proved.

2.3. For any $\mathcal{O}$-module $\mathcal{F}$ there exists an essential extension $\mathcal{F} \subset \mathcal{G}$ with injective $\mathcal{O}$-module $\mathcal{G}$. If $\mathcal{F} \subset \mathcal{G}'$ is another such extension, then there exists an isomorphism $\mathscr{G} \to \mathscr{G}'$, inducing the identity isomorphism of the $\mathcal{O}$-module $\mathscr{F}$ onto itself.

In fact, let $\mathscr{L}$ be an injective $\mathcal{O}$-module, of which $\mathscr{F}$ is a submodule. We denote by $\mathfrak{S}$ the set of all $\mathcal{O}$-modules $\mathscr{S} \subset \mathscr{L}$, for which the extension $\mathscr{F} \subset \mathscr{S}$ is essential. We order the set $\mathfrak{S}$ by inclusion. Obviously it is injective. By Zorn's lemma $\mathfrak{S}$ contains a maximum element $\mathscr{G}$, where $\mathscr{F} \subset \mathscr{G}$ is an essential extension of $\mathcal{O}$-modules. If $\mathscr{G} \subset \mathscr{G}_1$ is an essential extension of the module $\mathscr{G}$, then in view of the injectivity of $\mathscr{L}$ one can assume that $\mathscr{G}_1 \subset \mathscr{L}$, and then $\mathscr{G}_1 = \mathscr{G}$ by virtue of the maximality of $\mathscr{G}$. Thus, by Proposition 2.2, the $\mathcal{O}$-module $\mathscr{G}$ is injective.

If $\mathscr{F} \subset \mathscr{G}'$ is another essential extension with an injective $\mathcal{O}$-module $\mathscr{G}'$, then the diagram with exact row

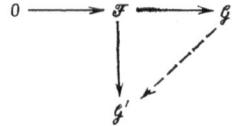

can be completed by the dashed arrow to a commutative diagram. From this one easily derives (cf. the proof of Proposition 2.2) that the map $\mathscr{G} \to \mathscr{G}'$ is an isomorphism, which induces the identity isomorphism of the $\mathcal{O}$-module $\mathscr{F}$ onto itself.

The essential extension of $\mathcal{O}$-modules $\mathscr{F} \subset \mathscr{G}$ with injective $\mathcal{O}$-module $\mathscr{G}$, unique up to equivalence, is called the *injective hull* of the $\mathcal{O}$-module $\mathscr{F}$. On the one hand, the $\mathcal{O}$-module $\mathscr{G}$ so defined is a minimal injective extension of the $\mathcal{O}$-module $\mathscr{F}$ and, on the other, is its maximal essential extension.[12]

2.4. Let $\mathcal{O}$ be a sheaf of fields on the topological space X. Then an extension $\mathscr{F} \subset \mathscr{G}$ of sheaves of vector spaces on $\mathcal{O}$ is essential if and only if, for each open set $U \subset X$ and each nonzero section $g \in \Gamma(U; \mathscr{G})$, there exists a point $x \in U$, such that $g_x \neq 0$ and $g_x \in \mathscr{F}_x$.

A sheaf of vector spaces $\mathscr{F}$ on $\mathcal{O}$ is injective if and only if it is a flabby sheaf. In fact, the condition of injectivity of the $\mathcal{O}$-module $\mathscr{F}$ is equivalent to the following: for any sheaf of ideals $\mathscr{I}$ in $\mathcal{O}$ the following sequence must be exact:

$$\mathrm{Hom}_6(X;\ 6,\ \mathcal{F}) \to \mathrm{Hom}_6(X;\ \mathcal{I},\ \mathcal{F}) \to 0.$$

Since 6 is a sheaf of fields, each subsheaf of ideals $\mathcal{I} \subset 6$ has the form $\mathcal{I} = 6_U$, where U is an open set in X. By virtue of the natural isomorphism

$$\mathrm{Hom}_6(X;\ 6_U,\ \mathcal{F}) = \Gamma(U;\ \mathcal{F})$$

the condition of injectivity of the 6-module $\mathcal{F}$ means that the restriction map

$$\Gamma(X;\ \mathcal{F}) \to \Gamma(U;\ \mathcal{F})$$

is surjective, i.e., that the sheaf $\mathcal{F}$ is flabby.

By Proposition 2.2 the 6-module $\mathcal{F}$ is flabby, if and only if it has no proper essential extensions. The latter means that for each extension $\mathcal{F} \subset \mathcal{G}$ which is proper there exists an open set $U \subset X$ and a nonzero section $g \in \Gamma(U; \mathcal{G})$ such that for each $x \in U$ either $g_x = 0$ or $g_x \notin \mathcal{F}_x$.

According to point 2.3, for any sheaf of vector spaces $\mathcal{F}$ on 6 there exists an essential extension $\mathcal{F} \subset \mathcal{G}$ with flabby sheaf of vector spaces $\mathcal{G}$ on 6, which is unique up to equivalence. It is natural to call the sheaf $\mathcal{G}$ the *flabby hull* of the sheaf $\mathcal{F}$.

2.5. Lemma. Let $\mathcal{F}$ be a flabby sheaf of Abelian groups on the topological spaces X. Then for each proper extension of sheaves of Abelian groups $\mathcal{F} \subset \mathcal{G}$ there exists an open set $U \subset X$ and a nonzero section $g \in \Gamma(U; \mathcal{G})$ such that, for each $x \in U$, either $g_x = 0$ or $g_x \notin \mathcal{F}_x$.

Proof. Since by hypothesis $\mathcal{F} \neq \mathcal{G}$, there exists an open set $U \subset X$ and a section $g_0 \in \Gamma(U; \mathcal{G})$, not belonging to the group $\Gamma(U; \mathcal{F})$. We denote by U_0 the set of those points $x \in U$ for which $g_{0x} \in \mathcal{F}_x$. Obviously U_0 is an open set, which does not coincide with U. The restriction of the section g_0 to U_0 belongs to the group $\Gamma(U_0; \mathcal{F})$. Since $\mathcal{F}$ is a flabby sheaf, there exists a section $f \in \Gamma(u; \mathcal{F})$, which coincides with g_0 on the set U_0. Then the section $g = g_0 - f \in \Gamma(u; \mathcal{G})$ satisfies the following conditions: $g_x = 0$ for $x \in U_0$ and $g_x \notin \mathcal{F}_x$ for $x \in U \setminus U_0$. The lemma is proved.

2.6. Let us agree on the following notation. Let 6 be a sheaf of commutative rings with units on the topological spaces X. For an arbitrary 6-module $\mathcal{F}$ and an arbitrary sheaf of ideals $\mathcal{I}$ in 6 let $\mathcal{I}\mathcal{F}$ denote the largest 6-submodule $\mathcal{M} \subset \mathcal{F}$ with respect to inclusion, satisfying the condition $\mathcal{I}\mathcal{M} = 0$. In other words,

$$_\mathcal{I}\mathcal{F} = \mathcal{H}om_\mathcal{O}\,(\mathcal{O}/\mathcal{I},\ \mathcal{F}).$$

If S is an arbitrary closed subset of X, then let $_S\mathcal{F}$ denote the sheaf on X which is defined by the presheaf $U \to \Gamma_S(U;\ \mathcal{F})$. In other words, $_S\mathcal{F} = \mathcal{H}_S^0(\mathcal{F})$ is the sheaf of germs of zero-dimensional local cohomology of the sheaf $\mathcal{F}$ with respect to the closed set S. It is also obvious that $_S\mathcal{F} = _\mathcal{I}\mathcal{F}$, where $\mathcal{I} = \mathcal{O}_{X\setminus S}$.

Lemma. Let $\mathcal{F}$ be an injective $\mathcal{O}$-module. Then for an arbitrary sheaf of ideals $\mathcal{I}$ in $\mathcal{O}$ the sheaf $_\mathcal{I}\mathcal{F}$ is an injective $\mathcal{O}/\mathcal{I}$ module.

Proof. Let $\mathcal{Y}$ be an arbitrary sheaf of ideals in $\mathcal{O}$, containing the sheaf $\mathcal{I}$. From the exact sequence

$$0 \to \mathcal{Y}/\mathcal{I} \to \mathcal{O}/\mathcal{I}$$

we get a commutative diagram

$$\begin{array}{ccccc}
\mathrm{Hom}_{\mathcal{O}/\mathcal{I}}\,(X;\ \mathcal{O}/\mathcal{I},\ _\mathcal{I}\mathcal{F}) & \to & \mathrm{Hom}_{\mathcal{O}/\mathcal{I}}\,(X;\ \mathcal{Y}/\mathcal{I},\ _\mathcal{I}\mathcal{F}) & \to & 0 \\
\downarrow & & \downarrow & & \\
\mathrm{Hom}_{\mathcal{O}}\,(X;\ \mathcal{O}/\mathcal{I},\ \mathcal{F}) & \to & \mathrm{Hom}_{\mathcal{O}}\,(X;\ \mathcal{Y}/\mathcal{I},\ \mathcal{F}) & & \to\ 0
\end{array}$$

in which the maps represented by vertical arrows are bijective and the lower row is exact by virtue of the injectivity of the $\mathcal{O}$-module $\mathcal{F}$. The injectivity of the $\mathcal{O}/\mathcal{I}$ module $_\mathcal{I}\mathcal{F}$ follows from the exactness of the upper row. The lemma is proved.

Corollary. Let $\mathcal{F}$ be an injective $\mathcal{O}$-module. Then for an arbitrary closed set $S \subset X$ the sheaf $_S\mathcal{F}$ is also an injective $\mathcal{O}$-module.

Proof. For an arbitrary $\mathcal{O}$-module $\mathcal{E}$ the natural map

$$\mathrm{Hom}_{\mathcal{O}_S}\,(X;\ \mathcal{E}_S,\ _S\mathcal{F}) \to \mathrm{Hom}_\mathcal{O}\,(X;\ \mathcal{E},\ _S\mathcal{F})$$

is bijective. By the preceding lemma, the sheaf $_S\mathcal{F}$ is an injective $\mathcal{O}_S$-module. Moreover, the functor $\mathcal{E} \to \mathcal{E}_S$ is exact. The assertion follows from this.

2.7. Lemma. Suppose, by induction, for each $k = 1, 2, \ldots$, there are given a point z_k in the space $\mathbf{C}^n$, and open set $U_k \subset \mathbf{C}^n$ and a finite family $F_k = (f_{ki})$ of holomorphic functions in U_k of the following form:

a) the point z_k belongs to the open set U_{k-1};

b) the family of holomorphic functions F_k is defined in a neighborhood of the point z_k and contains the family F_{k-1};

c) the set U_k is an open neighborhood of the point z_k, contained in U_{k-1} and in the domain of definition of F_k.

Suppose for each $k = 1, 2, \ldots$ the neighborhood U_k of the point z_k is chosen sufficiently small, and let $\mathcal{O}$ denote the sheaf of germs of holomorphic functions in the space $\mathbf{C}^n$. Then there exists an integer k_0 such that for each $k \geq k_0$ the family F_k is contained in the ideal of the ring $\Gamma(U_k; \mathcal{O})$ generated by the family F_{k0}.[13]

Proof. If the assertion of this lemma is true, then by induction on $m = 1, 2, \ldots$, we get that it is also true when f_{ki} are holomorphic vector-functions with values in the space $\mathbf{C}^m$, i.e., when $f_{ki} \in \Gamma(U_k; \mathcal{O}^m)$. We shall prove the assertion of the lemma by induction on n. We choose a coordinate system $(\zeta_1, \ldots, \zeta_n)$ in the space $\mathbf{C}^u$ so that the function f_{11} is regular with respect to ζ_n at the point z_1. By the Weierstrass preparation theorem, in a sufficiently small neighborhood U_1 of the point z_1 the function f_{11} can be represented in the form $f_{11} = uh$, where u is an invertible element of the ring $\Gamma(U_1; \mathcal{O})$, and h is a Weierstrass polynomial of some degree p with respect to $(\zeta_n - z_{1n})$. It follows from this that the function f_{11} is regular of order $\leq p$ with respect to ζ_n at each point $z \in U_1$. By the Weierstrass division theorem for each $k = 1, 2, \ldots$ one can choose the neighborhood U_k of the point z_k sufficiently small that the functions f_{ki} can be represented uniquely in the form

$$f_{ki} = g_{ki}f_{11} + r_{ki};$$

here $g_{ki} \in \Gamma(U_k; \mathcal{O})$, and r_{ki} is a polynomial of the form

$$r_{ki} = a_{ki1}(\zeta_n - z_{kn})^{p-1} + \ldots + a_{kip},$$

where a_{kij} are holomorphic functions with respect to the variables $\zeta_1, \ldots, \zeta_{n-1}$. Considering finite families of vector-functions $a_{ki} = (a_{ki1}, \ldots, a_{kip})$, by the inductive hypothesis we get the assertion of the lemma.[14]

2.8. Theorem. Let $\mathcal{O}$ be the sheaf of germs of holomorphic functions on the domain G of the space $\mathbf{C}^n$. Then the analytic sheaf $\mathcal{F}$ on G is injective as an $\mathcal{O}$-module if and only if the following conditions hold:

a) the sheaf $\mathscr{F}$ is flabby;

b) for any closed set $S \subset G$, for each point $z \in G$ the stalk $_S\mathscr{F}_z$ of the sheaf $_S\mathscr{F}$ is an injective $\mathcal{O}_z$-module.[15]

Proof. Let the $\mathcal{O}$-module $\mathscr{F}$ be injective. Then the conditions of the theorem hold. In fact, for an arbitrary open set $U \subset G$ from the exact sequence of sheaves $0 \to \mathcal{O}_U \to \mathcal{O}$ we get a commutative diagram

$$
\begin{array}{ccc}
\Gamma(G; \mathscr{F}) \to & \Gamma(U; \mathscr{F}) & \to 0 \\
\downarrow & \downarrow & \\
\mathrm{Hom}_{\mathcal{O}}(G; \mathcal{O}, \mathscr{F}) \to & \mathrm{Hom}_{\mathcal{O}}(G; \mathcal{O}_U, \mathscr{F}) & \to 0
\end{array}
$$

in which the lower horizontal row is exact, and the maps represented by vertical arrows are bijective. The exactness of the upper row means that $\mathscr{F}$ is a flabby sheaf. According to the corollary to Lemma 2.6, for an arbitrary closed set $S \subset G$ the sheaf $_S\mathscr{F}$ is an injective $\mathcal{O}$-module. Let I be an ideal in the ring $\mathcal{O}_z$. Then in some neighborhood of the point z there exists a coherent subsheaf of ideals $\mathscr{J} \subset \mathcal{O}$, for which $\mathscr{J}_z = I$. Since in a neighborhood of z there exists an exact sequence

$$\mathcal{O}^q \to \mathcal{O}^p \to \mathscr{J} \to 0,$$

there is a commutative diagram with exact rows

$$
\begin{array}{ccccccc}
0 \to & \mathscr{H}om_{\mathcal{O}}(\mathscr{J}, {}_S\mathscr{F})_z \to & \mathscr{H}om_{\mathcal{O}}(\mathcal{O}^p, {}_S\mathscr{F})_z \to & \mathscr{H}om_{\mathcal{O}}(\mathcal{O}^q, {}_S\mathscr{F})_z \\
& \downarrow & \downarrow & \downarrow \\
0 \to & \mathrm{Hom}_{\mathcal{O}_z}(I, {}_S\mathscr{F}_z) \to & \mathrm{Hom}_{\mathcal{O}_z}(\mathcal{O}_z^p, {}_S\mathscr{F}_z) \to & \mathrm{Hom}_{\mathcal{O}_z}(\mathcal{O}_z^q, {}_S\mathscr{F}_z)
\end{array}
$$

in which the maps represented by the second and third vertical arrows are bijective. Consequently, the map represented by the first vertical arrow is also bijective. From the exact sequence of sheaves $0 \to \mathscr{J} \to \mathcal{O}$, we get a commutative diagram

$$
\begin{array}{ccc}
\mathscr{H}om_{\mathcal{O}}(\mathcal{O}, {}_S\mathscr{F})_z \to & \mathscr{H}om_{\mathcal{O}}(\mathscr{J}, {}_S\mathscr{F})_z & \to 0 \\
\downarrow & \downarrow & \\
\mathrm{Hom}_{\mathcal{O}_z}(\mathcal{O}_z, {}_S\mathscr{F}_z) \to & \mathrm{Hom}_{\mathcal{O}_z}(I, {}_S\mathscr{F}_z) & \to 0
\end{array}
$$

in which the maps represented by vertical arrows are bijective and the upper horizontal row is exact by virtue of the injectivity of the $\mathcal{O}$-module $_S\mathscr{F}$. The exactness of the lower row means that $_S\mathscr{F}_z$ is an injective $\mathcal{O}_z$-module.

Let us now assume that the θ-module $\mathscr{F}$ satisfies the hypotheses of the theorem but is not injective. Then by Proposition 2.2 there exists a proper essential extension of θ-modules $\mathscr{F} \subset \mathscr{G}$. Since $\mathscr{F}$ is a flabby sheaf, by Lemma 2.5 there exists a nonempty open set $U \subset G$ and a nonzero section $g \in \Gamma(U; \mathscr{G})$ such that for each $z \in U$ either $g_z = 0$ or $g_z \notin \mathscr{F}_z$. By Lemma 2.1 there exists a point $z_1 \in U$ and a germ of a holomorphic function $\varphi_{1z_1} \in \theta_z$, such that $\varphi_{1z_1} g_{z_1} \neq 0$ and $\varphi_{1z_1} g_{z_1} \in \mathscr{F}_{z_1}$. We denote by S the support of the section g in U and we choose a sufficiently small open neighborhood U_1 of the point z_1 in U. Then there exists a section $f \in \Gamma_S(U_1; \mathscr{F})$, such that $\varphi_{1z_1} f_{z_1} = \varphi_{1z_1} g_{z_1}$. We set $g_1 = g - f$ in U_1. Then $g_{1z} = 0$ for $z \in U_1 \setminus S$ and $g_{1z} \notin \mathscr{F}_z$ for $z \in S \cap U_1$. Since $z_1 \in S$, then $S \cap U_1 \neq \varnothing$. Moreover, one can assume that φ_{1z_1} is the germ at the point z_1 of a holomorphic function $\varphi_1 \in \Gamma(U_1; \theta)$, such that $\varphi_1 g_1 = 0$ in U_1. Now continuing this process by induction, for each $k = 1, 2, \ldots$ we define a point $z_k \in G$, an open set $U_k \subset G$, a holomorphic function $\varphi_k \in \Gamma(U_k; \theta)$, and a section $g_k \in \Gamma(U_k; \mathscr{G})$, such that the following conditions hold:

(1) U_k is a sufficiently small open neighborhood of the point z_k contained in U_{k-1};

(2) the germ $(\varphi_k g_{k-1})_{z_k}$ is not equal to zero and belongs to the stalk $_S \mathscr{F}_{z_k}$;

(3) the intersection $S \cap U_k$ is not empty, and $g_{kz} = 0$ for $z \in U_k \setminus S$ and $g_{kz} \notin \mathscr{F}_z$ for $z \in S \cap U_k$.

(4) $\varphi_j g_k = 0$ for $j = 1, \ldots, k$ in U_k.

Let us assume that for each $j \leq k$ the point z_j, the open set U_j, the holomorphic function φ_j, and the section g_j are already defined so that (1)-(4) hold for them. It follows from (3) that the section $g_k \in \Gamma_S(U_k; \mathscr{G})$ is different from zero. By Lemma 2.1 there exists a point $z_{k+1} \in U_k$ and a holomorphic function φ_{k+1} in a neighborhood of this point, such that the germ of the product $(\varphi_{k+1} g_k)_{z_{k+1}}$ is not equal to zero and belongs to the stalk $_S \mathscr{F}_{z_{k+1}}$. Obviously $z_{k+1} \in S$, so S intersects each neighborhood of the point z_{k+1}. Since by the hypothesis of the theorem the stalk $_S \mathscr{F}_{z_{k+1}}$ is an injective $\theta_{z_{k+1}}$-module, over some sufficiently small neighborhood of the point z_{k+1} there exists a section f of the sheaf $_S \mathscr{F}$, satisfying the condition

$$(\varphi_j g_k)_{z_{k+1}} = (\varphi_j f)_{z_{k+1}} \quad \text{for} \quad j \leqslant k+1.$$

We choose a sufficiently small open neighborhood U_{k+1} of the point z_{k+1} and in this neighborhood we set $g_{k+1} = g_k - f$. Then it is obvious that $g_{(k+1)z} = 0$ for $z \in U_{k+1} \setminus S$ and $g_{(k+1)z} \notin \mathscr{F}_z$ for $z \in S \cap U_{k+1}$, while the intersection $S \cap U_{k+1}$ is not empty. Finally, $\varphi_j g_{k+1} = 0$ for $j \leq k + 1$, since the neighborhood U_{k+1} is sufficiently small. Thus the possibility of an inductive construction is proved. Conditions (2) and (4) mean that for each $k \geq 2$ in the open set U_k one has

$$\varphi_j g_{k-1} = 0 \quad (j \leqslant k-1); \quad \varphi_k g_{k-1} \neq 0.$$

Hence, for any $k = 2, 3, \ldots$, the function φ_k cannot belong to the ideal generated by the functions $\varphi_1, \ldots, \varphi_{k-1}$ in the ring $\Gamma(U_k; 6)$. In view of Lemma 2.7, we arrive at a contradiction. The theorem is proved.[16]

3. Local Duality[17]

3.1. Let X be a complex manifold of dimension n, which is countable at infinity, 6 be the sheaf of germs of holomorphic functions on X, $\mathscr{F}$ and $\mathscr{G}$ be coherent analytic sheaves on X, admitting free resolutions

$$0 \longrightarrow 6^{s_n} \xrightarrow{\sigma_n} \ldots \longrightarrow 6^{s_1} \xrightarrow{\sigma_1} 6^{s_0} \xrightarrow{\sigma_0} \mathscr{F} \longrightarrow 0;$$
$$0 \longrightarrow 6^{t_n} \xrightarrow{\tau_n} \ldots \longrightarrow 6^{t_1} \xrightarrow{\tau_1} 6^{t_0} \xrightarrow{\tau_0} \mathscr{G} \longrightarrow 0. \tag{1}$$

We consider the double complex

$$K^{p,q}(X) = \prod_{i-j=p} \mathrm{Hom}_6(X; 6^{s_i}, 6^{t_j} \otimes_6 \mathscr{E}^{0,q}),$$

whose differentials are defined by the homomorphisms σ_i, τ_j and the differential d'' of the Dolbeault–Grothendieck resolution

$$0 \longrightarrow 6 \longrightarrow \mathscr{E}^{0,0} \xrightarrow{d''} \mathscr{E}^{0,1} \xrightarrow{d''} \ldots \xrightarrow{d''} \mathscr{E}^{0,n} \longrightarrow 0$$

(cf. Dolbeault [1]; Gunning and Rossi [1]), where $\mathscr{E}^{0,q}$ is the sheaf of germs of C^∞ differential forms of double degree $(0, q)$ on X. We denote by $K^\bullet(X)$ the associated single complex

$$K^r(X) = \prod_{p+q=r} K^{p,q}(X).$$

Lemma. There exist natural isomorphisms of vector spaces

$$\operatorname{Ext}^k_{\mathcal{O}}(X; \mathcal{F}, \mathcal{G}) = H^k K^{\cdot}(X).$$

Proof. We consider the first filtration of the complex $K^{\cdot}(X)$ and we calculate the initial terms of the spectral sequence $E_r{}^{p,q}(X)$, corresponding to this filtration. Since there is a natural isomorphism

$$K^{p,q}(X) = \prod_{i-j=p} \Gamma(X; \mathcal{H}om_{\mathcal{O}}(\mathcal{O}^{s_i}, \mathcal{O}^{t_j}) \otimes_{\mathcal{O}} \mathcal{O}\mathcal{E}^{0,q}),$$

we get

$$E_1^{p,q}(X) = \prod_{i-j=p} H^q(X; \mathcal{H}om_{\mathcal{O}}(\mathcal{O}^{s_i}, \mathcal{O}^{t_j})).$$

Since the filtration is obviously regular, we get a convergent spectral sequence

$$E_2^{p,q}(X) \Rightarrow H^{p+q}K^{\cdot}(X).$$

Now we choose an injective resolution in the sense of Cartan–Eilenberg for the complex $\mathcal{O}^{t_{\cdot}}$:

$$0 \longrightarrow \mathcal{O}^{t_{\cdot}} \longrightarrow \mathcal{L}^{\cdot, 0} \longrightarrow \mathcal{L}^{\cdot, 1} \longrightarrow \cdots$$

(cf. Cartan and Eilenberg [1, Chapter XVII, Section 1, p. 434]). We consider the double complex

$$K^{p,q} = \prod_{i-j=p} \operatorname{Hom}_{\mathcal{O}}(X; \mathcal{O}^{s_i}, \mathcal{L}^{j,q})$$

and we denote by $K^{\cdot}$ the associated single complex. For the first filtration of the complex $K^{\cdot}$ we get

$${}'E_1^{p,q} = \prod_{i-j=p} \operatorname{Ext}^q_{\mathcal{O}}(X; \mathcal{O}^{s_i}, \mathcal{O}^{t_j}).$$

Since the first filtration is regular, we get a convergent spectral sequence

$${}'E_2^{p,q} \Rightarrow H^{p+q}K^{\cdot}.$$

Since the Cartan–Eilenberg resolution is injective, there exists a homomorphism of double complexes

$$\mathcal{O}^{t_j} \otimes_{\mathcal{O}} \mathcal{E}^{0,q} \longrightarrow \mathcal{L}^{j,q},$$

which is unique up to homotopy, and which in turn induces a homomorphism of double complexes

$$K^{p,\,q}(X) \to K^{p,\,q}.$$

The last homomorphism induces an isomorphism of the initial terms of the spectral sequences corresponding to the first filtrations, i.e., $E_1^{p,q}(X) = {}'E_1^{p,q}$. Consequently, we get a natural isomorphism

$$H^k K^{\cdot}(X) = H^k K^{\cdot}. \tag{2}$$

Now we consider the second filtration of the complex $K^{\cdot}$. For the corresponding spectral sequence we get

$${}''E_1^{p,\,q} = \operatorname{Hom}_{\mathcal{O}}(X;\ \mathcal{F},\ H_{-q}\mathcal{L}^{\cdot,\,p}).$$

Since for each j we have an injective resolution

$$0 \longrightarrow H_j \mathcal{O}^{\prime\cdot} \longrightarrow H_j \mathcal{L}^{\cdot,\,0} \longrightarrow H_j \mathcal{L}^{\cdot,\,1} \longrightarrow \ldots,$$

and $H_j \mathcal{O}^{\prime\cdot} = 0$ for $j \neq 0$ and $H_0 \mathcal{O}^{\prime\cdot} = \mathcal{G}$, the spectral sequence considered degenerates:

$$\begin{aligned}
{}''E_2^{p,\,q} &= 0 \quad \text{for} \quad q \neq 0; \\
{}''E_2^{p,\,0} &= \operatorname{Ext}_{\mathcal{O}}^p(X;\ \mathcal{F},\ \mathcal{G}).
\end{aligned} \tag{3}$$

Since the second filtration of the complex $K^{\cdot}$ is regular, there is a natural isomorphism

$$H^k K^{\cdot} = {}''E_2^{k,\,0}.$$

Comparing this isomorphism with the isomorphisms (2) and (3), we get the assertion of the lemma.

3.2. We consider the double complex

$$K_c^{p,\,q}(X) = \prod_{j-i=p} \operatorname{Hom}_{\mathcal{O},\,c}(X;\ \mathcal{O}^{\prime j},\ \mathcal{O}^{s i} \otimes_{\mathcal{O}} {}'\mathcal{D}^{n,\,q}),$$

whose differential is defined by the homomorphisms σ_i, τ_j of the free resolutions (1) and the differential d'' of the Dolbeault–Grothendieck resolution

$$0 \longrightarrow \Omega^n \longrightarrow {}'\mathcal{D}^{n,\,0} \xrightarrow{d''} {}'\mathcal{D}^{n,\,1} \xrightarrow{d''} \ldots \xrightarrow{d''} {}'\mathcal{D}^{n,\,n} \longrightarrow 0$$

(cf. Dolbeault [1]; Schwartz [2, p. 64]), where Ω^n is the sheaf of germs of holomorphic forms of degree n and ${}'\mathcal{D}^{n,q}$ is the sheaf of germs of currents of double degree $(n,\ q)$ on X. We denote by $K_c^{\cdot}(X)$ the associated single complex

$$K^r_c(X) = \prod_{p+q=r} K^{p,\,q}_c(X).$$

Lemma. There exist natural isomorphisms of vector spaces

$$\operatorname{Ext}^k_{\mathcal{O},\,c}(X;\,\mathscr{G},\,\mathscr{F}\otimes_{\mathcal{O}}\Omega^n) = H^k K^{\bullet}_c(X).$$

The **proof** is analogous to the proof of Lemma 3.1.

3.3. Lemma. Let $U\colon E\to F$ and $v\colon F\to G$ be continuous linear maps of Frechet–Schwartz spaces such that $v\circ u = 0$. Then there exists a natural isomorphism of topological vector spaces

$$\{\operatorname{Ker} v/\overline{\operatorname{Im} u}\}' = \operatorname{Ker} {}^t u/\overline{\operatorname{Im} {}^t v},$$

where ${}^t u\colon F'\to E'$ and ${}^t v\colon G'\to F'$ are the dual maps, and the dual spaces considered are endowed with the strong topology.[18]

Proof. There exists a natural isomorphism of topological vector spaces

$$\operatorname{Ker} {}^t u = \{F/\overline{\operatorname{Im} u}\}'.$$

Applying the Hahn–Banach theorem, we get a surjective homomorphism of the strong duals to Frechet–Schwartz spaces

$$\operatorname{Ker} {}^t u \longrightarrow \{\operatorname{Ker} v/\overline{\operatorname{Im} u}\}',$$

whose kernel coincides with $\overline{\operatorname{Im} {}^t v}$. The assertion of the lemma follows directly from this.

3.4. We endow the vector space $K^{p,q}(X)$ with its natural Frechet–Schwartz space topology. In the vector space $\operatorname{Ext}_{\mathcal{O}}{}^k(X;\,\mathscr{F},\mathscr{G})$ we define a topology with the help of the isomorphism of Lemma 3.1. Then the associated separated space $\widetilde{\operatorname{Ext}}_{\mathcal{O}}{}^k(X;\,\mathscr{F},\mathscr{G})$ is a Frechet–Schwartz space (the tilde here denotes factorization by the closure of zero). We endow the vector space $K_c{}^{p,q}(X)$ with its natural topology as strong dual to a Frechet–Schwartz space. In the vector space $\operatorname{Ext}_{\mathcal{O},c}{}^k(X;\,\mathscr{G},\,\mathscr{F}\otimes_{\mathcal{O}}\Omega^n)$ we define a topology with the help of the isomorphism of Lemma 3.2. Then the associated separated space $\widetilde{\operatorname{Ext}}_{\mathcal{O},c}{}^k(X;\,\mathscr{G},\,\mathscr{F}\otimes_{\mathcal{O}}\Omega^n)$ is the strong dual to a Frechet–Schwartz space.

Theorem. The topological vector space $\widetilde{\mathrm{Ext}}_{\mathcal{O},c}{}^k(X; \mathcal{G}, \mathcal{F} \otimes_{\mathcal{O}} \Omega^n)$ is naturally isomorphic to the strong dual to the topological vector space $\widetilde{\mathrm{Ext}}_{\mathcal{O}}{}^{n-k}(X; \mathcal{F}, \mathcal{G})$: $\{ \widetilde{\mathrm{Ext}}\,_{\mathcal{O}}^{k}(X; \mathcal{F}, \mathcal{G}) \}' = \widetilde{\mathrm{Ext}}\,_{\mathcal{O},\,c}^{n-k}(X; \mathcal{G}, \mathcal{F} \otimes_{\mathcal{O}} \Omega^n)$.

Proof. The strong dual to the topological vector space $\Gamma(X; \mathcal{H}om_{\mathcal{O}}(\mathcal{O}^{s_i}, \mathcal{O}^{t_j}) \otimes_{\mathcal{O}} \mathcal{S}^{0,q})$ is naturally isomorphic to the topological vector space $\Gamma_c(X; \mathcal{H}om_{\mathcal{O}}(\mathcal{O}^{t_j}, \mathcal{O}^{s_i}) \otimes_{\mathcal{O}} '\mathcal{D}^{n,n-q})$. Consequently, the strong dual to the topological vector space $K^{p,q}(X)$ is naturally isomorphic to the topological vector space $K_c^{-p,n-q}(X)$. Further, the strong dual to the topological vector space $K^r(X)$ is naturally isomorphic to the topological vector space $K_c^{n-r}(X)$. Finally, the map dual to the differential $K^r(X) \to K^{r+1}(X)$ of the complex $K^\cdot(X)$ coincides up to sign with the differential $K_c^{n-r-1}(X) \to K_c^{n-r}(X)$ of the complex $K_c^\cdot(X)$. Thus, the assertion of the theorem follows from Lemma 3.3.

3.5.   Corollaries.

3.5.1. The topological vector space $\mathrm{Ext}_{\mathcal{O}}{}^k(X; \mathcal{F}, \mathcal{G})$ is separated if and only if the topological vector space $\mathrm{Ext}_{\mathcal{O},c}{}^{n-k+1}(X; \mathcal{G}, \mathcal{F} \otimes_{\mathcal{O}} \Omega^n)$ is separated.

3.5.2. (Serre–Malgrange Duality). There exist natural isomorphisms of topological vector spaces[19]

$$\{ \tilde{H}^k(X; \mathcal{F}) \}' = \widetilde{\mathrm{Ext}}\,_{\mathcal{O},\,c}^{n-k}(X; \mathcal{F}, \Omega^n);$$
$$\{ \tilde{H}_c^k(X; \mathcal{F}) \}' = \widetilde{\mathrm{Ext}}\,_{\mathcal{O}}^{n-k}(X; \mathcal{F}, \Omega^n).$$

Proof. The first isomorphism is obtained from Theorem 3.4 for $\mathcal{F} = \mathcal{O}$, the second for $\mathcal{G} = \Omega^n$.

3.5.3. If the manifold X is holomorphically complete, then

$$H_c^k(X; \mathcal{O}) = 0 \quad \text{for} \quad k \neq n$$

and there exists a natural isomorphism of topological vector spaces

$$H_c^n(X; \mathcal{O}) = \{ \Gamma(X; \Omega^n) \}'.$$

Proof. From the first isomorphism of Corollary 3.5.2 we get the isomorphism

$$\{\tilde{H}^k(X;\ \Omega^n)\}' = \tilde{H}_c^{n-k}(X;\ \theta).$$

The assertion now follows from Cartan's theorem (B) (point 1.13) and Corollary 3.5.1.

3.6. Lemma. Let $E^{\cdot}$ and $F^{\cdot}$ be complexes in the category of Frechet spaces (or in the category of strong duals to Frechet–Schwartz spaces) with continuous linear differentials. Let $u\colon E^{\cdot} \to F^{\cdot}$ be a continuous linear morphism of complexes, which is an algebraic quasiisomorphism, i.e., it induces an isomorphism of cohomology topological vector spaces[20]

$$H^k E^{\cdot} = H^k F^{\cdot}$$

Proof. Since the map of cohomology topological vector spaces $H^k E^{\cdot} \to H^k F^{\cdot}$ induced by the morphism u is bijective and continuous, it suffices to prove that it maps the closure of zero to the closure of zero. The map

$$u + d\colon Z^k E^{\cdot} \bigoplus F^{k-1} \to Z^k F^{\cdot}$$

is surjective and, consequently, by Banach's theorem, is a homomorphism of topological vector spaces. The set $\overline{dE^{k-1}} \oplus F^{k-1}$ is saturated with respect to the map $u + d$. Consequently, its image is closed in $Z^k F^{\cdot}$, and hence coincides with $\overline{dF^{k-1}}$. The lemma is proved.

3.7. Theorem. If the complex manifold X is holomorphically complete, then the topological vector space $\mathrm{Ext}_{\theta, c}^{n-k}(X;\ \mathcal{G},\ \mathcal{F} \otimes_{\theta} \Omega^n)$ is separated and is naturally isomorphic to the strong dual to the separated topological vector space $\mathrm{Ext}_{\theta}^{k}(X;\ \mathcal{F}, \mathcal{G})$.

Proof. It suffices to prove the separation of the topological vector space $\mathrm{Ext}_{\theta}^{k}(X;\ \mathcal{F}, \mathcal{G})$, since the assertion then follows from Theorem 3.4 and Corollary 3.5.1. We consider the double complex

$$K^{p,\,q} = \mathrm{Hom}_{\theta}(X;\ \theta^{s_p},\ \theta^{t-q})$$

and the associated single complex

$$K^r = \prod_{p+q=r} K^{p,\,q}.$$

By Cartan's theorem (B) (point 1.13)

$$\operatorname{Ext}_{\mathcal{O}}^{k}(X;\, \mathcal{O}^{s_{P}},\, \mathcal{G}) = 0 \quad \text{for} \quad k \neq 0.$$

Consequently, for the first filtration of the complex $K^{\cdot}$ we get a spectral sequence in which

$$'E_{1}^{p,\,q} = 0 \quad \text{for} \quad q \neq 0;$$
$$'E_{1}^{p,\,0} = \operatorname{Hom}_{\mathcal{O}}(X;\, \mathcal{O}^{s_{P}},\, \mathcal{G}).$$

Thus the spectral sequence degenerates:

$$'E_{2}^{p,\,q} = 0 \quad \text{for} \quad q \neq 0;$$
$$'E_{2}^{p,\,0} = \operatorname{Ext}_{\mathcal{O}}^{p}(X;\, \mathcal{F},\, \mathcal{G}).$$

Since the first filtration is regular, we get a natural isomorphism of vector spaces

$$\operatorname{Ext}_{\mathcal{O}}^{k}(X;\, \mathcal{F},\, \mathcal{G}) = H^{k}K^{\cdot}.$$

We use the spectral sequence

$$E_{2}^{p,\,q} = H^{p}(X;\, \mathscr{E}\mathrm{xt}_{\mathcal{O}}^{q}(\mathcal{F},\, \mathcal{G})) \Rightarrow \operatorname{Ext}_{\mathcal{O}}^{p+q}(X;\, \mathcal{F},\, \mathcal{G})$$

(cf. point 1.10.2). By Cartan's theorem (B) (point 1.13) this spectral sequence degenerates. We get a natural isomorphism of vector spaces

$$\operatorname{Ext}_{\mathcal{O}}^{k}(X;\, \mathcal{F},\, \mathcal{G}) = \Gamma(X;\, \mathscr{E}\mathrm{xt}_{\mathcal{O}}^{k}(\mathcal{F},\, \mathcal{G})).$$

Consequently, the natural continuous linear map

$$H^{k}K^{\cdot} \to \Gamma(X;\, \mathscr{E}\mathrm{xt}_{\mathcal{O}}^{k}(\mathcal{F},\, \mathcal{G}))$$

is bijective, so the topological vector space $H^{k}K^{\cdot}$ is separated. By what was proved above, the inclusion map $K^{r} \to K^{r}(X)$ is an algebraic quasiisomorphism of complexes in the category of Frechet–Schwartz spaces. To complete the proof of the theorem it suffices to apply Lemma 3.6 now. The theorem is proved.

3.8. Theorem. If the complex manifold X is compact, then the vector space $\operatorname{Ext}_{\mathcal{O}}^{n-k}(X;\, \mathcal{G},\, \mathcal{F} \otimes_{\mathcal{O}} \Omega^{n})$ is finite-dimensional and naturally isomorphic to the algebraic dual to the finite-dimensional vector space $\operatorname{Ext}_{\mathcal{O}}^{k}(X;\, \mathcal{F},\, \mathcal{G})$.

Proof. We use the spectral sequence

$$E_2^{p,\,q} = H^p(X;\ \mathcal{E}\mathrm{xt}_{\mathcal{O}}^q(\mathcal{F},\,\mathcal{G})) \Rightarrow \mathrm{Ext}_{\mathcal{O}}^{p+q}(X;\ \mathcal{F},\,\mathcal{G})$$

(point 1.10.2). By the Cartan–Serre finiteness theorem (point 1.15.2) the vector spaces $E_2{}^{p,q}$ are finite-dimensional. Consequently, the topological vector space $\mathrm{Ext}_{\mathcal{O}}{}^k(X;\ \mathcal{F},\mathcal{G})$ is finite-dimensional and separated. The assertion now follows from Theorem 3.4 and Corollary 3.5.1.[21]

3.9. Let S be an arbitrary closed set in the complex manifold X. For fixed j and q we define the cone of the cochain mapping

$$\mathrm{Hom}_{\mathcal{O}}(X;\ \mathcal{O}^{s_i},\ \mathcal{O}^{t_j}\otimes_{\mathcal{O}}\mathcal{O}^{0,\,q}) \to \mathrm{Hom}_{\mathcal{O}}(X\setminus S;\ \mathcal{O}^{s_i},\mathcal{O}^{t_j}\otimes_{\mathcal{O}}\mathcal{O}^{0,\,q}),$$

by defining its components and differentials in the familiar way (cf. MacLane [1, pp. 67–68]). We consider the double complex

$$K_S^{p,\,q}(X) = \prod_{i-j=p} \mathrm{Hom}_{\mathcal{O}}(X;\ \mathcal{O}^{s_i},\ \mathcal{O}^{t_j}\otimes_{\mathcal{O}}\mathcal{O}^{0,\,q}) \times \mathrm{Hom}_{\mathcal{O}}(X\setminus S;\mathcal{O}^{s_i-1},\mathcal{O}^{t_j}\otimes_{\mathcal{O}}\mathcal{O}^{0,\,q}),$$

and also the associated single complex

$$K_S^r(X) = \prod_{p+q=r} K_S^{p,\,q}(X).$$

Lemma. There exist natural isomorphisms of vector spaces

$$\mathrm{Ext}_{\mathcal{O},\,S}^k(X;\ \mathcal{F},\,\mathcal{G}) = H^k K_S^{\bullet}(X).$$

Proof. For the first filtration of the complex $K_S{}^{\bullet}(X)$ we get a convergent spectral sequence

$$E_{S,\,1}^{p,\,q}(X) \Rightarrow H^{p+q} K_S^{\bullet}(X),$$

in which

$$E_{S,\,1}^{p,\,q}(X) = \prod_{i-j=p} H^q(X;\ \mathcal{H}\mathrm{om}_{\mathcal{O}}(\mathcal{O}^{s_i},\ \mathcal{O}^{t_j})) \times H^q(X\setminus S;\ \mathcal{H}\mathrm{om}_{\mathcal{O}}(\mathcal{O}^{s_i-1},\mathcal{O}^{t_j})).$$

For fixed j and q we define the cone of the cochain map

$$\mathrm{Hom}_{\mathcal{O}}(X;\ \mathcal{O}^{s_i},\ \mathscr{L}^{j,\,q}) \to \mathrm{Hom}_{\mathcal{O}}(X\setminus S;\mathcal{O}^{s_i},\mathscr{L}^{j,\,q}),$$

where $\mathscr{L}^{j,q}$ are the components of the Cartan–Eilenberg resolution

$$0 \to \mathcal{O}^{t_\bullet} \to \mathscr{L}^{\bullet,\,0} \to \mathscr{L}^{\bullet,\,1} \to \cdots,$$

and we consider the double complex

$$K^{p,\,q} = \prod_{i-j=p} \mathrm{Hom}_{\mathcal{O}}(X;\, \mathcal{O}^{s_i},\, \mathcal{L}^{j,\,q}) \times \mathrm{Hom}_{\mathcal{O}}(X\backslash S;\, \mathcal{O}^{s_{i-1}},\, \mathcal{L}^{j,\,q}),$$

and also the associated single complex

$$K^r = \prod_{p+q=r} K^{p,\,q}.$$

For the first filtration of the complex $K^{\cdot}$ we get a convergent spectral sequence

$$'E_2^{p,\,q} \Rightarrow H^{p+q} K^{\cdot},$$

in which

$$'E_1^{p,\,q} = \prod_{i-j=p} \mathrm{Ext}^q_{\mathcal{O}}(X;\, \mathcal{O}^{s_i},\, \mathcal{O}^{t_j}) \times \mathrm{Ext}^q_{\mathcal{O}}(X\backslash S;\, \mathcal{O}^{s_{i-1}},\, \mathcal{O}^{t_j}).$$

Thus, there exists a homomorphism of double complexes

$$K_S^{p,\,q}(X) \to K^{p,\,q},$$

which induces an isomorphism $E_{S,1}{}^{p,q} = {}'E_1{}^{p,q}$. Since the filtrations are regular, we get a natural isomorphism

$$H^k K_S^{\cdot}(X) = H^k K^{\cdot}. \tag{4}$$

For the second filtration of the complex $K^{\cdot}$ we get a spectral sequence in which

$$''E_1^{p,\,q} = \mathrm{Hom}_{\mathcal{O},\,S}(X;\, \mathcal{F},\, H_{-q}\mathcal{L}^{\cdot,\,p}).$$

Just as in the proof of Lemma 3.1, we get that the spectral sequence considered degenerates:

$$''E_2^{p,\,q} = 0 \quad \text{for} \quad q \neq 0; \tag{5}$$
$$''E_2^{p,\,0} = \mathrm{Ext}^p_{\mathcal{O},\,S}(X;\, \mathcal{F},\, \mathcal{G}).$$

The second filtration of the complex $K^{\cdot}$ is also regular, so there is a natural isomorphism

$$H^k K^{\cdot} = {}''E_2^{k,\,0}.$$

The assertion of the lemma now follows from (4) and (5). The lemma is proved.

3.10. We define for fixed j and q the cone of the cochain map

$$\text{Hom}_{\mathcal{O}, \, c}(X \backslash S; \, \mathcal{O}^{tj}, \, \mathcal{O}^{si} \otimes_{\mathcal{O}}' \mathcal{D}^{n, \, q}) \rightarrow \text{Hom}_{\mathcal{O}, \, c}(X; \, \mathcal{O}^{tj}, \, \mathcal{O}^{si} \otimes_{\mathcal{O}}' \mathcal{D}^{n, \, q})$$

and we consider the double complex

$$K_c^{p, \, q}(S) = \prod_{j - i = p} \text{Hom}_{\mathcal{O}, \, c}(X; \, \mathcal{O}^{tj}, \, \mathcal{O}^{si} \otimes_{\mathcal{O}}' \mathcal{D}^{n, \, q})$$
$$\times \text{Hom}_{\mathcal{O}, \, c}(X \backslash S; \, \mathcal{O}^{tj}, \, \mathcal{O}^{si-1} \otimes_{\mathcal{O}}' \mathcal{D}^{n, \, q}),$$

and also the single complex associated with it

$$K_c^r(S) = \prod_{p + q = r} K_c^{p, \, q}(S).$$

Lemma. There exist natural isomorphisms of vector spaces $\text{Ext}_{\mathcal{O}, c}^k(S; \mathcal{G}, \mathcal{F} \otimes_{\mathcal{O}} \Omega^n, = H^k K_c^{\cdot}(S)$.

The proof is analogous to the proof of Lemma 3.9.

3.11. We endow the vector space $K_{\mathcal{S}}^{p, q}(X)$ with its natural topology as a Frechet–Schwartz space. Thus there is defined a topology on the vector space $K_{\mathcal{S}}^r(X)$ which will also be a Frechet–Schwartz space. On the vector space $\text{Ext}_{\mathcal{O}, \mathcal{S}}^k(X; \mathcal{F}, \mathcal{G})$ we introduce a topology by means of the isomorphism of Lemma 3.9. Then the associated separated space $\widetilde{\text{Ext}}_{\mathcal{O}, \mathcal{S}}^k(X; \mathcal{F}, \mathcal{G})$ will be a Frechet–Schwartz space. We endow the vector space $K_c^{p, q}(S)$ with its natural topology as the strong dual to a Frechet–Schwartz space. This defines a topology on the vector space $K_c^r(S)$. On the vector space $\text{Ext}_{\mathcal{O}, c}^k(S; \mathcal{G}, \mathcal{F} \otimes_{\mathcal{O}} \Omega^n)$ we introduce a topology by means of the isomorphism of Lemma 3.10. Then the associated separated space $\widetilde{\text{Ext}}_{\mathcal{O}, c}^k(S; \mathcal{G}, \mathcal{F} \otimes_{\mathcal{O}} \Omega^n)$ will be the strong dual to a Frechet–Schwartz space.

Theorem. The topological vector space $\widetilde{\text{Ext}}_{\mathcal{O}, c}^k(S; \mathcal{G}, \mathcal{F} \otimes_{\mathcal{O}} \Omega^n)$ is naturally isomorphic to the strong dual to the topological vector space $\widetilde{\text{Ext}}_{\mathcal{O}, \mathcal{S}}^{n-k}(X; \mathcal{F}, \mathcal{G})$:

$$\{\widetilde{\text{Ext}}_{\mathcal{O}, \mathcal{S}}^k(X; \mathcal{F}, \mathcal{G})\}' = \widetilde{\text{Ext}}_{\mathcal{O}, c}^{n-k}(S; \mathcal{G}, \mathcal{F} \otimes_{\mathcal{O}} \Omega^n).$$

Proof. The strong dual to the topological vector space $\text{Hom}_{\mathcal{O}}(X; \mathcal{O}^{si}, \mathcal{O}^{tj} \otimes_{\mathcal{O}} \mathcal{E}^{0, q})$ is naturally isomorphic to the topological vector space

$$\text{Hom}_{6,\,c}\,(X;\; 6^t{}_j,\; 6^s{}_i \otimes {}_6'\mathscr{D}^{n,\,n-q}).$$

Hence the strong dual to the topological vector space $K_S{}^{p,q}(X)$ is naturally isomorphic to the topological vector space $K_c{}^{-p,n-q}(S)$. Consequently, the strong dual to the topological vector space $K_S{}^r(X)$ is naturally isomorphic to the topological vector space $K_c{}^{n-r}(S)$. The map dual to the differential $K_S{}^r(X) \to K_S{}^{r+1}(X)$ of the complex $K_S{}^{\cdot}(X)$ coincides up to sign with the differential $K_c{}^{n-r-1}(S) \to K_c{}^{n-r}(S)$ of the complex $K_c{}^{\cdot}(S)$. Thus, the assertion of the theorem follows from Lemma 3.3. The theorem is proved.

3.12. Corollaries.

3.12.1. The topological vector space $\text{Ext}_{6,S}{}^k(X;\,\mathscr{F},\mathscr{G})$ is separated if and only if the topological vector space $\text{Ext}_{6,c}{}^{n-k+1}(S;\,\mathscr{G},\,\mathscr{F}\otimes_6\Omega^n)$ is separated.

3.12.2. There exist natural isomorphisms of topological vector spaces

$$\{\tilde{H}_S^k(X;\,\mathscr{F})\}' = \widetilde{\text{Ext}}_{6,\,c}^{\,n-k}(S;\,\mathscr{F},\,\Omega^n);$$
$$\{\tilde{H}_c^k(S;\,\mathscr{F})\}' = \widetilde{\text{Ext}}_{6,\,S}^{\,n-k}(X;\,\mathscr{F},\,\Omega^n).$$

Proof. The first isomorphism is obtained from Theorem 3.11 for $\mathscr{F} = 6$, the second, for $\mathscr{G} = \Omega^n$.

3.12.3. Let S be a compact set in X, which has a fundamental system of holomorphically complete open neighborhoods. Then

$$\text{Ext}_{6,\,S}^{\,k}(X;\,\mathscr{F},\,\Omega^n) = 0 \quad \text{for} \quad k \neq n;$$
$$\text{Ext}_{6,\,S}^{\,n}(X;\,\mathscr{F},\,\Omega^n) = \{\Gamma(S;\,\mathscr{F})\}'.$$

In particular, setting $\mathscr{F} = \Omega^n$, we get[22]

$$H_S^k(X;\,6) = 0 \quad \text{for} \quad k \neq n;$$
$$H_S^n(X;\,6) = \{\Gamma(S;\,\Omega^n)\}'.$$

3.13. Theorem. If S is a compact set in the complex manifold X, which has a fundamental system of holomorphically complete open neighborhoods, then the topological vector space $\text{Ext}_6{}^{n-k}(S;\,\mathscr{G},\,\mathscr{F}\otimes_6\Omega^n)$ is separated and naturally isomorphic to the strong dual to the separated topological vector space $\text{Ext}_{6,S}{}^k(X;\,\mathscr{F},\mathscr{G})$.

Proof. We consider the double complex of topological vector spaces

$$K^{p,q} = \text{Hom}_{\mathcal{O}}(S; \; \mathcal{O}^{t_p}, \; \mathcal{O}^{s-q} \otimes_{\mathcal{O}} \Omega^n)$$

and the associated single complex

$$K^r = \coprod_{p+q=r} K^{p,q}$$

By Cartan's theorem (B) (point 1.13)

$$\text{Ext}_{\mathcal{O}}^k(S; \; \mathcal{O}^{t_p}, \; \mathcal{F} \otimes_{\mathcal{O}} \Omega^n) = 0 \quad \text{for} \quad k \neq 0.$$

Consequently, for the first filtration of the complex $K^{\cdot}$ we get a spectral sequence in which

$$'E_1^{p,q} = 0 \quad \text{for} \quad q \neq 0,$$
$$'E_1^{p,0} = \text{Hom}_{\mathcal{O}}(S; \; \mathcal{O}^{t_p}, \; \mathcal{F} \otimes_{\mathcal{O}} \Omega^n).$$

Thus, the spectral sequence degenerates:

$$'E_2^{p,q} = 0 \quad \text{for} \quad q \neq 0,$$
$$'E_2^{p,0} = \text{Ext}_{\mathcal{O}}^p(S; \; \mathcal{G}, \; \mathcal{F} \otimes_{\mathcal{O}} \Omega^n).$$

Since the filtration considered is regular, we get a natural isomorphism of vector spaces

$$\text{Ext}_{\mathcal{O}}^k(S; \; \mathcal{G}, \; \mathcal{F} \otimes_{\mathcal{O}} \Omega^n) = H^k K^{\cdot}. \tag{6}$$

Now we make use of the spectral sequence

$$E_2^{p,q} = H^p(S; \; \mathcal{E}xt_{\mathcal{O}}^q(\mathcal{G}, \; \mathcal{F} \otimes_{\mathcal{O}} \Omega^n)) \Rightarrow \text{Ext}_{\mathcal{O}}^{p+q}(S; \; \mathcal{G}, \; \mathcal{F} \otimes_{\mathcal{O}} \Omega^n).$$

By Cartan's Theorem (B) it degenerates. We get a natural isomorphism of vector spaces

$$\text{Ext}_{\mathcal{O}}^k(S; \; \mathcal{G}, \; \mathcal{F} \otimes_{\mathcal{O}} \Omega^n) = \Gamma(S; \; \mathcal{E}xt_{\mathcal{O}}^k(\mathcal{G}, \; \mathcal{F} \otimes_{\mathcal{O}} \Omega^n)).$$

Thus, the natural continuous linear map

$$H^k K^{\cdot} \longrightarrow \Gamma(S; \; \mathcal{E}xt_{\mathcal{O}}^k(\mathcal{G}, \; \mathcal{F} \otimes_{\mathcal{O}} \Omega^n))$$

is bijective. Consequently, the topological vector space $H^k K^{\cdot}$ is separated. It follows from (6) that the inclusion map

$$K^r \longrightarrow K^r_c(S)$$

is an algebraic quasiisomorphism of complexes in the category of strong duals to Frechet–Schwartz spaces. The assertion now follows from Lemma 3.6. The theorem is proved.

4. Injective and Global Dimension

4.1. Let $\mathcal{O}$ be a sheaf of commutative rings with unit on the topological space X. By *the injective dimension of the $\mathcal{O}$-module $\mathcal{F}$* is meant the smallest integer $m \geq 0$, having the property that there exists an exact sequence of the form

$$0 \longrightarrow \mathcal{F} \longrightarrow \mathcal{L}^0 \longrightarrow \mathcal{L}^1 \longrightarrow \cdots \longrightarrow \mathcal{L}^m \longrightarrow 0,$$

in which the $\mathcal{O}$-modules $\mathcal{L}^0, \mathcal{L}^1, ..., \mathcal{L}^m$ are injective (i.e., $\mathcal{L}^{\cdot}$ is an injective resolution of the $\mathcal{O}$-module $\mathcal{F}$ of length m). The injective dimension of the $\mathcal{O}$-module $\mathcal{F}$ is denoted by inj dim$_\mathcal{O} \mathcal{F}$. It is considered to be infinite if the $\mathcal{O}$-module $\mathcal{F}$ does not have any injective resolutions of finite length. In other words, the injective dimension of the $\mathcal{O}$-module $\mathcal{F}$ is either infinite or the smallest integer m such that

$$\operatorname{Ext}^k_\mathcal{O}(X; \mathcal{E}, \mathcal{F}) = 0 \quad \text{for} \quad k > m$$

for any $\mathcal{O}$-module $\mathcal{E}$.

Lemma. The property inj dim$_\mathcal{O} \mathcal{F} \leq m$ is local, i.e., it holds if and only if inj dim$_{\mathcal{O}|U} \mathcal{F}|U \leq m$ for each sufficiently small open set $U \subset X$.

Proof. Let $\mathcal{L}^{\cdot}$ be a resolution of the $\mathcal{O}$-module $\mathcal{F}$ of length m, in which the $\mathcal{O}$-modules $\mathcal{L}^0, \mathcal{L}^1, ..., \mathcal{L}^{m-1}$ are injective. If inj dim$_{\mathcal{O}|U} \mathcal{F}|U \leq m$ for each sufficiently small open set $U \subset X$, then the $\mathcal{O} | U$-module $\mathcal{L}^m |U$ is injective. It follows from this that the $\mathcal{O}$-module $\mathcal{L}^m$ has no proper essential extensions, i.e., by Proposition 2.2 it is injective. The latter means that the injective dimension of the $\mathcal{O}$-module $\mathcal{F}$ does not exceed m. Conversely, if inj dim$_\mathcal{O} \mathcal{F} \leq m$, then there exists an injective resolution $\mathcal{L}^{\cdot}$ of length m of the $\mathcal{O}$-module $\mathcal{F}$. Consequently, for an arbitrary open set $U \subset X$ we get an injective resolution of the $\mathcal{O} | U$-module $\mathcal{F}|U$

$$0 \longrightarrow \mathcal{F} | U \longrightarrow \mathcal{L}^0 | U \longrightarrow \cdots \longrightarrow \mathcal{L}^m | U \longrightarrow 0$$

of length m. This means that $\operatorname{inj\,dim}_{\theta\,|U}\mathscr{F}\,|\,U \le m$. The lemma is proved.

4.2. Theorem. Let θ be the sheaf of germs of holomorphic functions in the domain G of the space $\mathbb{C}^n$. Then the analytic sheaf $\mathscr{F}$ on G has the property $\operatorname{inj\,dim}_{\theta}\mathscr{F} \le m$, if and only if the following conditions hold:

a) for each open set $U \subset G$ the natural map

$$H^m(G;\ \mathscr{F}) \longrightarrow H^m(U;\ \mathscr{F})$$

is surjective;

b) for each coherent subsheaf of ideals $\mathscr{I} \subset \theta$ and each closed set $S \subset G$ the natural map

$$\mathscr{H}^m_S(\mathscr{F}) = \mathscr{E}\mathrm{xt}^m_{\theta,\,S}(\theta,\ \mathscr{F}) \longrightarrow \mathscr{E}\mathrm{xt}^m_{\theta,\,S}(\mathscr{I},\ \mathscr{F})$$

is surjective.

Proof. Suppose given an exact sequence

$$0 \longrightarrow \mathscr{F} \longrightarrow \mathscr{L}^0 \longrightarrow \mathscr{L}^1 \longrightarrow \cdots \longrightarrow \mathscr{L}^m \longrightarrow 0,$$

in which the θ-modules $\mathscr{L}^0, \mathscr{L}^1, \ldots, \mathscr{L}^{m-1}$ are injective. It suffices to prove that the injectivity of the θ-module $\mathscr{L}^m$ follows from the conditions of the theorem. We consider the commutative diagram with exact rows

$$
\begin{array}{ccccccc}
\Gamma(G;\ \mathscr{L}^{m-1}) & \longrightarrow & \Gamma(G;\ \mathscr{L}^m) & \longrightarrow & H^m(G;\ \mathscr{F}) & \longrightarrow & 0 \\
\downarrow & & \downarrow & & \downarrow & & \\
\Gamma(U;\ \mathscr{L}^{m-1}) & \longrightarrow & \Gamma(U;\ \mathscr{L}^m) & \longrightarrow & H^m(U;\ \mathscr{F}) & \longrightarrow & 0
\end{array}
$$

Since the extreme vertical arrows are surjective maps, according to the five lemma the middle vertical arrow

$$\Gamma(G;\ \mathscr{L}^m) \longrightarrow \Gamma(U;\ \mathscr{L}^m)$$

is also a surjective map. Consequently, $\mathscr{L}^m$ is a flabby sheaf. Analogously, we consider the commutative diagram with exact rows

$$
\begin{array}{ccccccc}
\mathscr{H}\mathrm{om}_{\theta,\,S}(\theta, \mathscr{L}^{m-1}) & \longrightarrow & \mathscr{H}\mathrm{om}_{\theta,\,S}(\theta, \mathscr{L}^m) & \longrightarrow & \mathscr{E}\mathrm{xt}^m_{\theta,\,S}(\theta, \mathscr{F}) & \longrightarrow & 0 \\
\downarrow & & \downarrow & & \downarrow & & \\
\mathscr{H}\mathrm{om}_{\theta,\,S}(\mathscr{I}, \mathscr{L}^{m-1}) & \longrightarrow & \mathscr{H}\mathrm{om}_{\theta,\,S}(\mathscr{I}, \mathscr{L}^m) & \longrightarrow & \mathscr{E}\mathrm{xt}^m_{\theta,\,S}(\mathscr{I}, \mathscr{F}) & \longrightarrow & 0
\end{array}
$$

in which $\mathcal{I}$ is a coherent subsheaf of ideals in $\mathcal{O}$, and S is a closed subset of G. Since the extreme vertical arrows are surjective maps, the middle vertical map

$$\mathcal{H}\text{om}_{\mathcal{O}, S}(\mathcal{O}, \mathcal{L}^m) \to \mathcal{H}\text{om}_{\mathcal{O}, S}(\mathcal{I}, \mathcal{L}^m)$$

is also surjective. By virtue of the natural isomorphism

$$\mathcal{H}\text{om}_{\mathcal{O}, S}(\mathcal{I}, \mathcal{L}^m) = \mathcal{H}\text{om}_{\mathcal{O}}(\mathcal{I}, {}_S\mathcal{L}^m)$$

this means that for each $z \in G$ the stalk ${}_S\mathcal{L}_z^m$ of the sheaf ${}_S\mathcal{L}^m$ (cf. point 2.6) is an injective $\mathcal{O}_z$-module. Thus, by Theorem 2.8, the $\mathcal{O}$-module $\mathcal{L}^m$ is injective. The theorem is proved.

4.3. Corollary. Let $\mathcal{O}$ be a sheaf of germs of holomorphic functions in the domain G of the space $\mathbf{C}^n$. Then the analytic sheaf $\mathcal{F}$ on G has the property inj dim $_{\mathcal{O}} \mathcal{F} \leq m$ if and only if

$$\text{Ext}_{\mathcal{O}, S}^k(G; \mathcal{O}/\mathcal{I}, \mathcal{F}) = 0 \quad \text{for} \quad k > m$$

for each coherent subsheaf of ideals $\mathcal{I} \subset \mathcal{O}$ and each closed subset $S \subset G$.

Proof. The necessity of this condition follows directly from the natural isomorphism

$$\text{Ext}_{\mathcal{O}, S}^k(G; \mathcal{O}/\mathcal{I}, \mathcal{F}) = \text{Ext}_{\mathcal{O}}^k(G; (\mathcal{O}/\mathcal{I})_S, \mathcal{F}).$$

Let us assume that the condition of Corollary 4.3 holds. Then, in particular,

$$H_S^k(G; \mathcal{F}) = 0 \quad \text{for} \quad k > m.$$

Consequently, condition a) of Theorem 4.2 holds. For an arbitrary open set $U \subset G$ we consider the exact sequence

$$\ldots \to \text{Ext}_{\mathcal{O}, S}^k(G; \mathcal{O}/\mathcal{I}, \mathcal{F}) \to \text{Ext}_{\mathcal{O}, S}^k(U; \mathcal{O}/\mathcal{I}, \mathcal{F}) \to$$
$$\to \text{Ext}_{\mathcal{O}, S \setminus U}^{k+1}(G; \mathcal{O}/\mathcal{I}, \mathcal{F}) \to \ldots$$

Then we get

$$\text{Ext}_{\mathcal{O}, S}^k(U; \mathcal{O}/\mathcal{I}, \mathcal{F}) = 0 \quad \text{for} \quad k > m.$$

Since the open set U is arbitrary, for $k > m$ we get $\mathscr{E}xt_{\mathfrak{G}, S}^k(\mathfrak{G}/\mathfrak{I}, \mathscr{F}) = 0$, i.e., condition b) of Theorem 4.2 holds. The corollary is proved.

4.4. Theorem. Let $\mathfrak{G}$ be the sheaf of germs of holomorphic functions in the domain G of the space $\mathbf{C}^n$. Then for any nonzero coherent analytic sheaf $\mathscr{F}$ on G

$$\text{inj dim}_{\mathfrak{G}} \mathscr{F} = n.$$

Proof. First we prove the inequality $\text{inj dim}_{\mathfrak{G}} \mathscr{F} \leq n$. Let $\mathscr{I}$ be a coherent subsheaf of ideals in $\mathfrak{G}$ and S be a closed subset in G. Then

$$\text{Ext}_{\mathfrak{G}, c}^k(S; \mathscr{F}, \mathfrak{G}/\mathfrak{I} \otimes_{\mathfrak{G}} \Omega^n) = 0 \quad \text{for} \quad k < 0.$$

The natural map

$$\text{Hom}_{\mathfrak{G}, c}(S; \mathscr{F}, \mathfrak{G}/\mathfrak{I} \otimes_{\mathfrak{G}} \Omega^n) \rightarrow \prod_{z \in S} \text{Hom}_{\mathfrak{G}}(\{z\}; \mathscr{F}, \mathfrak{G}/\mathfrak{I} \otimes_{\mathfrak{G}} \Omega^n)$$

is injective and by Lemma 3.10 continuous. By Theorem 3.13 the topological vector spaces

$$\text{Ext}_{\mathfrak{G}}^k(\{z\}; \mathscr{F}, \mathfrak{G}/\mathfrak{I} \otimes_{\mathfrak{G}} \Omega^n) = \text{Ext}_{\mathfrak{G}_z}^k(\mathscr{F}_z, \mathfrak{G}_z/\mathfrak{I}_z \otimes_{\mathfrak{G}_z} \Omega_z^n)$$

are separated. Consequently, the topological vector space $\text{Hom}_{\mathfrak{G}, c}(S; \mathscr{F}, \mathfrak{G}/\mathfrak{I} \otimes_{\mathfrak{G}} \Omega^n)$ is separated. By Corollary 3.12.1 the space $\text{Ext}_{\mathfrak{G}, S}^{n-k+1}(G; \mathfrak{G}/\mathfrak{I}, \mathscr{F})$ is separated for $k \leq 0$. By Theorem 3.11 we get

$$\text{Ext}_{\mathfrak{G}, S}^k(G; \mathfrak{G}/\mathfrak{I}, \mathscr{F}) = 0 \quad \text{for} \quad k > n.$$

By Corollary 4.3 the inequality $\text{inj dim}_{\mathfrak{G}} \mathscr{F} \leq n$ is proved. Now we prove the opposite inequality $\text{inj dim}_{\mathfrak{G}} \mathscr{F} \geq n$. Let $z \in G$ be a fixed point at which $\mathscr{F}_z \neq 0$. We denote by $\mathscr{I} = m(z)$ the subsheaf of ideals in $\mathfrak{G}$, consisting of germs of holomorphic functions equal to zero at the point z. Then the topological vector spaces

$$\text{Ext}_{\mathfrak{G}, c}^k(G; \mathscr{F}, \mathfrak{G}/\mathfrak{I} \otimes_{\mathfrak{G}} \Omega^n) = \text{Ext}_{\mathfrak{G}}^k(\{z\}; \mathscr{F}, \mathfrak{G}/\mathfrak{I} \otimes_{\mathfrak{G}} \Omega^n)$$

are separated by Theorem 3.13. The space

$$\text{Hom}_{\mathfrak{G}, c}(G; \mathscr{F}, \mathfrak{G}/\mathfrak{I} \otimes_{\mathfrak{G}} \Omega^n) = \text{Hom}_{\mathfrak{G}_z}(\mathscr{F}_z, \Omega_z^n/\mathfrak{I}_z \Omega_z^n)$$

is different from zero, since it is isomorphic to the space $\mathbf{C}^r$, where r is the minimal number of generators of the $\mathfrak{G}_z$-module $\mathscr{F}_z$. In fact, let $\{s_1, \ldots, s_r\}$ be a minimal system of generators in $\mathscr{F}_z$ (i.e., it is impossible to omit any of the $s_1, \ldots, s_r$). For arbitrary $\xi_1, \ldots, \xi_r \in \Omega_z^n / \mathcal{I}_z \Omega_z^n$ we set $f(s_i) = \xi_i$ $(1 \le i \le r)$. We get a homomorphism

$$f: \mathcal{F}_z \longrightarrow \Omega_z^n / \mathcal{I}_z \Omega_z^n,$$

since if $\varphi_1 s_1 + \ldots + \varphi_r s_r = 0$ for some $\varphi_1, \ldots, \varphi_r \in \mathcal{O}_z$, then $\varphi_1, \ldots, \varphi_r \in \mathcal{I}_z$ and, consequently, $\varphi_1 \xi_1 + \ldots + \varphi_r \xi_r = 0$. By Theorem 3.4 we get an isomorphism

$$\mathrm{Ext}_{\mathcal{O}}^n (G; \mathcal{O}/\mathcal{I}, \mathcal{F}) = \mathbf{C}^r.$$

By Corollary 4.3, inj dim $_{\mathcal{O}} \mathcal{F} \ge n$. The theorem is proved.[23]

4.5. Lemma. Let X be a locally compact topological space. Then for any sheaf of Abelian groups $\mathcal{F}$ on X there exists a spectral sequence

$$E_2^{p, \, q} = \lim{}^{(p)} H^q (K; \mathcal{F}) \Rightarrow H^{p+q} (X; \mathcal{F}),$$

where the derived functors of the projective limit are taken with respect to the filtered set of all compact subsets $K \subset X$, ordered by inclusion.

Proof. The set $\mathfrak{R}$ of compact subsets of the space X will be said to be saturated if for each $K \in \mathfrak{R}$ all closed subsets of the set K and also a neighborhood of K belong to the set $\mathfrak{R}$. If $\mathcal{L}$ is a flabby sheaf of Abelian groups on X, then the projective system $K \mapsto \Gamma(K; \mathcal{L})$ is "flabby" in the following sense: for any saturated sets $\mathfrak{R}' \subset \mathfrak{R}''$ of compact subsets of X the natural map

$$\lim_{\mathfrak{R}''} \Gamma (K; \mathcal{L}) \longrightarrow \lim_{\mathfrak{R}'} \Gamma (K; \mathcal{L})$$

is surjective. This is equivalent to the fact that for any open sets $U' \subset U''$ in X the restriction map $\Gamma(U''; \mathcal{L}) \to \Gamma(U'; \mathcal{L})$ is surjective. Thus, as is easy to show (cf. Roos [1]), the projective system $K \mapsto \Gamma(K; \mathcal{L})$ is acyclic:

$$\lim{}^{(p)} \Gamma (K; \mathcal{L}) = 0 \quad \text{for} \quad p \ge 1.$$

On the other hand, there is a natural isomorphism

$$\Gamma (X; \mathcal{L}) = \lim \Gamma (K; \mathcal{L}).$$

The assertion of the lemma follows from this (cf. Grothendieck [2, p. 50]).

4.6. Corollary (Milnor Exact Sequence). Let X be a locally compact topological space which is countable at infinity. Then for any sheaf of Abelian groups $\mathscr{F}$ on X there is an exact sequence

$$0 \longrightarrow \varprojlim{}^{(1)} H^{n-1}(K; \mathscr{F}) \longrightarrow H^n(X; \mathscr{F}) \longrightarrow \varprojlim H^n(K; \mathscr{F}) \longrightarrow 0,$$

where the projective limit functors are taken with respect to the filtered set, ordered by inclusion, of all compact subsets $K \subset X$.[24]

Proof. Under the assumptions made the filtered set, ordered by inclusion, of all compact subsets of X contains a countable cofinal subset. It follows from this that $\varprojlim^{(p)} = 0$ for $p \geq 2$. Hence in the spectral sequence of Lemma 4.5, $E_2^{p,q} = 0$ for $p \neq 0.1$. In this case, $E_r^{p,q} = 0$ for $p \neq 0.1$ and each $r \geq 2$. Consequently,

$$E_\infty^{p,\,q} = 0 \quad \text{for} \quad p \neq 0,1.$$

Thus the corollary is proved.

4.7. Lemma. Let X be a locally compact topological space, which is countable at infinity. For an arbitrary family $\mathfrak{B}$ of relatively compact open subsets of X and an arbitrary sheaf of Abelian groups $\mathscr{F}$ on X, we set

$$\mathscr{F}_{\mathfrak{B}} = \prod_{V \in \mathfrak{B}} \mathscr{F}_V.$$

Then the projective limit $\varprojlim H^n(K; \mathscr{F}_{\mathfrak{B}})$ with respect to the filtered set, ordered by inclusion, of all compact subsets $K \subset X$ can be identified naturally with a subgroup of the product

$$\prod_{V \in \mathfrak{B}} H_c^n(V; \mathscr{F}).$$

A family of cohomology classes $h_V \in H_c^n(V; \mathscr{F})(V \in \mathfrak{B})$ belongs to this subgroup if and only if the family of open sets $\{V \in \mathfrak{B}: h_V \neq 0\}$ is locally finite.

Proof. For an arbitrary compact set $K \subset X$ there is a natural isomorphism

$$H^n(K; \mathscr{F}_{\mathfrak{B}}) = \prod_{V \in \mathfrak{B}} H^n(K; \mathscr{F}_V)$$

(cf. Godement [1, p. 220]). Moreover, for an arbitrary open set $V \subset X$ there is a natural isomorphism

$$H^n(K; \mathscr{F}_V) = H^n_c(K \cap V; \mathscr{F}).$$

Since each set $V \in \mathfrak{B}$ is relatively compact, for it

$$H^n_c(V; \mathscr{F}) = \lim_{\leftarrow} H^n_c(K \cap V; \mathscr{F}).$$

Consequently, there exists a natural injective map

$$\lim_{\leftarrow} H^n(K; \mathscr{F}_{\mathfrak{B}}) \longrightarrow \prod_{V \in \mathfrak{B}} H^n_c(V; \mathscr{F}).$$

The assertion of the lemma follows directly from this.

4.8. Lemma. Let $\mathcal{O}$ be a coherent sheaf of Noetherian commutative rings with unit on the paracompact topological space X. Let $\mathscr{G}$ be a fine $\mathcal{O}$-module, for which for each $x \in X$ the stalk $\mathscr{G}_x$ is an injective $\mathcal{O}_x$-module. Then for an arbitrary coherent $\mathcal{O}$-module $\mathscr{F}$

$$\mathrm{Ext}^k_{\mathcal{O}}(X; \mathscr{F}, \mathscr{G}) = 0 \quad \text{for} \quad k \geqslant 1.$$

Proof. We make use of the spectral sequence

$$E_2^{p,q} = H^p(X; \mathscr{E}xt^q_{\mathcal{O}}(\mathscr{F}, \mathscr{G})) \Rightarrow \mathrm{Ext}^{p+q}_{\mathcal{O}}(X; \mathscr{F}, \mathscr{G})$$

(cf. point 1.10.2). Since $\mathscr{G}$ is a fine sheaf, the sheaves $\mathscr{E}xt_{\mathcal{O}}^q(\mathscr{F}, \mathscr{G})$ are fine. Consequently, the spectral sequence degenerates: $E_2^{p,q} = 0$ for $p \neq 0$. We get an isomorphism

$$\mathrm{Ext}^k_{\mathcal{O}}(X; \mathscr{F}, \mathscr{G}) = \Gamma(X; \mathscr{E}xt^k_{\mathcal{O}}(\mathscr{F}, \mathscr{G})).$$

On the other hand, under the assumptions made, for each point $x \in X$ there is a natural isomorphism

$$\mathscr{E}xt^k_{\mathcal{O}}(\mathscr{F}, \mathscr{G})_x = \mathrm{Ext}^k_{\mathcal{O}_x}(\mathscr{F}_x, \mathscr{G}_x)$$

(cf. Grothendieck [2, pp. 115–116]). From this we get that $\mathscr{E}xt_{\mathcal{O}}^k(\mathscr{F}, \mathscr{G}) = 0$ for $k \geq 1$. The lemma is proved.

4.8.1. Corollary. Let $\mathcal{O}$ be a coherent sheaf of Noetherian commutative rings with unit on the paracompact topological space X. Let

$$0 \longrightarrow \mathcal{G} \longrightarrow \mathcal{L}^0 \longrightarrow \mathcal{L}^1 \longrightarrow \cdots$$

be a resolution of the $\mathcal{O}$-module $\mathcal{G}$ consisting of fine $\mathcal{O}$-modules $\mathcal{L}^k$, for which for each $x \in X$ the stalk $\mathcal{L}_x^k$ is an injective $\mathcal{O}_x$-module. Then for an arbitrary coherent $\mathcal{O}$-module $\mathcal{F}$ there are natural isomorphisms

$$\mathrm{Ext}_{\mathcal{O}}^k(X; \ \mathcal{F}, \ \mathcal{G}) = H^k \mathrm{Hom}_{\mathcal{O}}(X; \ \mathcal{F}, \ \mathcal{L}^{\cdot}).$$

Proof. For each $n = 0, 1, \ldots$, let $\mathcal{Z}^n$ be the kernel of the homomorphism $\mathcal{L}^n \to \mathcal{L}^{n+1}$. We get an exact sequence

$$0 \longrightarrow \mathcal{Z}^n \longrightarrow \mathcal{L}^n \longrightarrow \mathcal{Z}^{n+1} \longrightarrow 0.$$

Thus the following sequence is exact:

$$\cdots \longrightarrow \mathrm{Ext}_{\mathcal{O}}^{k-1}(X; \ \mathcal{F}, \ \mathcal{L}^n) \longrightarrow \mathrm{Ext}_{\mathcal{O}}^{k-1}(X; \ \mathcal{F}, \ \mathcal{Z}^{n+1}) \longrightarrow$$
$$\longrightarrow \mathrm{Ext}_{\mathcal{O}}^k(X; \ \mathcal{F}, \ \mathcal{Z}^n) \longrightarrow \mathrm{Ext}_{\mathcal{O}}^k(X; \ \mathcal{F}, \ \mathcal{L}^n) \longrightarrow \cdots$$

From this, for $k \geq 2$ we get a natural isomorphism

$$\mathrm{Ext}_{\mathcal{O}}^k(X; \ \mathcal{F}, \ \mathcal{Z}^n) = \mathrm{Ext}_{\mathcal{O}}^{k-1}(X; \ \mathcal{F}, \ \mathcal{Z}^{n+1}).$$

Consequently, for $k \geq 1$,

$$\mathrm{Ext}_{\mathcal{O}}^k(X; \ \mathcal{F}, \ \mathcal{G}) = \mathrm{Ext}_{\mathcal{O}}^1(X; \ \mathcal{F}, \ \mathcal{Z}^{k-1}).$$

From the exact sequence $0 \to \mathcal{Z}^{k-1} \to \mathcal{L}^{k-1} \to \mathcal{Z}^k \to 0$ we get the exact sequence

$$\mathrm{Hom}_{\mathcal{O}}(X; \ \mathcal{F}, \ \mathcal{L}^{k-1}) \longrightarrow \mathrm{Hom}_{\mathcal{O}}(X; \ \mathcal{F}, \ \mathcal{Z}^k) \longrightarrow \mathrm{Ext}_{\mathcal{O}}^1(X; \mathcal{F}, \mathcal{Z}^{k-1}) \longrightarrow 0.$$

The assertion follows directly from this.

4.9. Let $\mathcal{O}$ be a sheaf of commutative rings with unit on the topological space X. By *the global dimension* of the sheaf $\mathcal{O}$ is meant the smallest integer m, which has the property that $\mathrm{inj\,dim}_{\mathcal{O}}\mathcal{F} \leq m$ for each $\mathcal{O}$-module $\mathcal{F}$. The global dimension of the sheaf of rings $\mathcal{O}$ is denoted by $\mathrm{gl\,dim}\,\mathcal{O}$. It is considered to be infinity if there exist $\mathcal{O}$-module of arbitrarily large injective dimension. The property $\mathrm{gl\,dim}\,\mathcal{O} \leq m$ is local, i.e., it holds if and only if $\mathrm{gl\,dim}\,\mathcal{O} \mid U \leq m$ for each sufficiently small open set $U \subset X$.

Theorem. Let 6 be the sheaf of germs of holomorphic functions on the domain G of the space $\mathbf{C}^n$, where $n \geq 1$. Then[25]

$$\text{gl dim } 6 = n + 1.$$

Proof. First we prove that gl dim $6 \geq n + 1$. Let S be an arbitrary nonempty closed set in G, for which the open set $U = G \setminus S$ is nonempty. Let $\mathfrak{V}$ be the family of all relatively compact holomorphically complete open sets in U. Let us assume that gl dim $6 \leq n$. Then, in particular,

$$H_S^{n+1}(G; \ 6_\mathfrak{V}) = 0.$$

Consequently, the restriction map

$$H^n(G; \ 6_\mathfrak{V}) \to H^n(U; \ 6_\mathfrak{V})$$

is surjective. On the other hand, there is a commutative diagram of Milnor exact sequences

$$0 \to \varprojlim{}^{(1)} H^{n-1}(K; \ 6_\mathfrak{V}) \to H^n(G; \ 6_\mathfrak{V}) \to \varprojlim \ H^n(K; \quad 6_\mathfrak{V}) \to 0$$
$$0 \to \varprojlim{}^{(1)} H^{n-1}(K|U; 6_\mathfrak{V}) \to H^n(U; 6_\mathfrak{V}) \to \varprojlim H^n(K|U; 6_\mathfrak{V}) \to 0$$

(cf. Corollary 4.6), where K and $K \mid U$ run through all compact subsets respectively in G and U. It follows directly from Lemma 4.7 that the right vertical map is not surjective. Hence the middle vertical map is also not surjective. We have found a contradiction.

4.9.1. Let 6 be the sheaf of germs of holomorphic functions on the domain G of the space $\mathbf{C}^n$, $\mathscr{F}$ be an arbitrary analytic sheaf on G, $\mathscr{I}$ be a coherent subsheaf of ideals in 6, S be a closed subset of G. We make use of the spectral sequence of point 1.10.2

$$E_2^{p, \ q} = H_S^p(G; \ \mathscr{E}\mathrm{xt}_6^q(6/\mathscr{I}, \ \mathscr{F})) \Rightarrow \mathrm{Ext}_{6, \ S}^{p+q}(G; \ 6/\mathscr{I}, \ \mathscr{F}).$$

Since the space $\mathbf{C}^n$ has real dimension $2n$, we get from the exact cohomology sequence associated with the open set $G \setminus S$ (cf. point (1.8.2)) that $E_2^{p,q} = 0$ for $p > 2n + 1$ (cf. Godement [1, p. 266]). On the other hand, by Hilbert's syzygy theorem (cf. Gunning and Rossi [1, p. 97]), in view of the natural isomorphism

$$\mathscr{E}\mathrm{xt}_6^q(6/\mathscr{I}, \ \mathscr{F})_z = \mathrm{Ext}_{6_z}^q(6_z/\mathscr{I}_z, \ \mathscr{F}_z)$$

we get that $E_2^{p,q} = 0$ for $q > n$. Consequently,

$$\mathrm{Ext}^k_{\mathcal{O},\,S}(G;\ \mathcal{O}/\mathfrak{I},\ \mathcal{F})=0 \quad \text{for} \quad k>3n+1.$$

From this, by virtue of Corollary 4.3, we get

$$\mathrm{gl\ dim}\ \mathcal{O} \leqslant 3n+1.$$

4.9.2. We show that gl dim $\mathcal{O} \leq n+1$. For this it suffices to prove that

$$\mathrm{Ext}^k_{\mathcal{O}}(G;\ \mathcal{F},\ \mathcal{G})=0 \quad \text{for} \quad k\geqslant n+2$$

for any analytic sheaves $\mathcal{F}$ and $\mathcal{G}$. For an arbitrary analytic sheaf $\mathcal{G}$ there exists a family $\mathfrak{B}$ of relatively compact holomorphically complete open sets in G such that sections of the sheaf $\mathcal{G}$ over sets of $\mathfrak{B}$ define an epimorphism of analytic sheaves $\mathcal{O}_{\mathfrak{B}} \to \mathcal{G}$. We denote by $\mathcal{R}$ the kernel of this epimorphism. From the exact sequence

$$0 \to \mathcal{R} \to \mathcal{O}_{\mathfrak{B}} \to \mathcal{G} \to 0$$

we get the exact sequence

$$\ldots \to \mathrm{Ext}^k_{\mathcal{O}}(G;\ \mathcal{F},\ \mathcal{O}_{\mathfrak{B}}) \to \mathrm{Ext}^k_{\mathcal{O}}(G;\ \mathcal{F},\ \mathcal{G}) \to \mathrm{Ext}^{k+1}_{\mathcal{O}}(G;\ \mathcal{F},\ \mathcal{R}) \to \ldots$$

In view of point 4.9.1, by induction on k we get that our assertion is valid if

$$\mathrm{Ext}^k_{\mathcal{O}}(G;\ \mathcal{F},\ \mathcal{O}_{\mathfrak{B}})=0 \quad \text{for} \quad k\geqslant n+2,$$

i.e., if inj dim $_{\mathcal{O}}\mathcal{O}_{\mathfrak{B}} \leq n+1$. Thus, according to Corollary 4.3 it suffices to prove that

$$\mathrm{Ext}^k_{\mathcal{O},\,S}(G;\ \mathcal{O}/\mathfrak{I},\ \mathcal{O}_{\mathfrak{B}})=0 \quad \text{for} \quad k\geqslant n+2$$

for an arbitrary coherent sheaf of ideals $\mathcal{I} \subset \mathcal{O}$ and an arbitrary closed set $S \subset G$. For this we make use of the exact sequence

$$\ldots \to \mathrm{Ext}^k_{\mathcal{O}}(G;\ \mathcal{O}/\mathfrak{I},\ \mathcal{O}_{\mathfrak{B}}) \to \mathrm{Ext}^k_{\mathcal{O}}(G\setminus S;\ \mathcal{O}/\mathfrak{I},\ \mathcal{O}_{\mathfrak{B}}) \to$$
$$\to \mathrm{Ext}^{k+1}_{\mathcal{O},\,S}(G;\ \mathcal{O}/\mathfrak{I},\ \mathcal{O}_{\mathfrak{B}}) \to \mathrm{Ext}^{k+1}_{\mathcal{O}}(G;\ \mathcal{O}/\mathfrak{I},\ \mathcal{O}_{\mathfrak{B}}) \to \ldots$$

It is clear from it that it suffices to prove the equality

$$\mathrm{Ext}^k_{\mathcal{O}}(U;\ \mathcal{O}/\mathfrak{I},\ \mathcal{O}_{\mathfrak{B}})=0 \quad \text{for} \quad k\geqslant n+1$$

for an arbitrary open set $U \subset G$. Let $\mathscr{L}$ be an injective analytic sheaf on G. Since there exists a homomorphism, left inverse to the inclusion homomorphism $\mathscr{L} \to \mathscr{C}^0(\mathscr{L})$, the sheaf $\mathscr{L}$ is fine. It follows from this that $\mathscr{L}_\mathfrak{B}$ is a fine sheaf having injective stalk. By Theorem 4.4 there exists an injective resolution

$$0 \to \mathscr{G} \to \mathscr{L}^0 \to \mathscr{L}^1 \to \ldots \to \mathscr{L}^n \to 0$$

of length n. From it we get the resolution

$$0 \to \mathscr{G}_\mathfrak{B} \to \mathscr{L}_\mathfrak{B}^0 \to \mathscr{L}_\mathfrak{B}^1 \to \ldots \to \mathscr{L}_\mathfrak{B}^n \to 0,$$

in which $\mathscr{L}_\mathfrak{B}{}^k$ are fine sheaves with injective stalks. From this, by Corollary 4.8.1 one gets a natural isomorphism

$$\operatorname{Ext}_\mathscr{G}^k (U; \ \mathscr{G}/\mathscr{I}, \ \mathscr{G}_\mathfrak{B}) = H^k \operatorname{Hom}_\mathscr{G} (U; \ \mathscr{G}/\mathscr{I}, \ \mathscr{L}_\mathfrak{B}^{\textstyle\cdot}),$$

from which our assertion follows directly. Theorem 4.9 is completely proved.[26]

5. Properties of Fine Sheaves

5.1. Let X be a locally compact topological space which is countable at infinity. We consider the set of all open subsets of X as a category, in which the set $\operatorname{Hom}(U, U')$ consists of one element for $U \subset U'$ and is empty for $U \subset U'$. By a *copresheaf* of Abelian groups on the category of all open sets in X is meant a covariant functor from this category to the category of Abelian groups. Analogously one defines a copresheaf on an arbitrary full subcategory of the category of all open sets in X. If, for example, $\mathscr{F}$ is a sheaf of Abelian groups on the topological space X, then the correspondence

$$\mathscr{F}_c \colon \ U \to \Gamma_c (U; \ \mathscr{F})$$

defines a copresheaf of Abelian groups on the category of all open sets in X.

Let $\mathfrak{B}$ be a basis of the topology in X consisting of relatively compact open sets and having the following property: for any two sets belonging to $\mathfrak{B}$, their intersection also belongs to $\mathfrak{B}$.. Let $\mathfrak{U} = (U_i)$ be a covering of the open set $U \subset X$ by open sets belonging to the basis $\mathfrak{B}$. For an arbitrary copresheaf F on the basis $\mathfrak{B}$ for each $k = 0, 1, \ldots$ we set

$$C_k^c (\mathfrak{U}; F) = \prod_{i_0, \ldots, i_k} F(U_{i_0 \ldots i_k}),$$

where the direct sum is taken over those collections of indices $i_0, \ldots, i_k$ for which the intersection $U_{i_0 \ldots i_k} = U_{i_0} \cap \ldots \cap U_{i_k}$ is not empty. We define the boundary operator

$$\partial \colon C_k^c (\mathfrak{U}; F) \to C_{k-1}^c (\mathfrak{U}; F),$$

by setting, for each $f = (f_{i_0 \ldots i_k}) \in C_k^c(\mathfrak{U}; F)$,

$$(\partial f)_{i_0 \ldots i_{k-1}} = \sum_{j} \sum_{s=0}^{k} (-1)^s f_{i_0 \ldots i_{s-1} j i_s \ldots i_{k-1}},$$

where j assumes only those values for which the intersection $U_j \cap U_{i_0 \ldots i_{k-1}}$ is not empty. Since $\partial \circ \partial = 0$, we get a chain complex $C \cdot{}^c(\mathfrak{U}; F)$, whose homology groups will be denoted by $H_k^c(\mathfrak{U}; F)$.[27]

5.2. We denote by $\mathfrak{B} \mid U$ the basis of open sets in the open set $U \subset X$ consisting of those sets of the basis $\mathfrak{B}$, whose closures are contained in U.

Lemma. Let $\mathscr{F}$ be a fine sheaf of Abelian groups on the topological space X. Then for any open set $U \subset X$ there is a natural isomorphism

$$H_k^c (\mathfrak{B} \mid U; \ \mathscr{F}_c) = 0 \quad \text{for} \quad k \neq 0,$$
$$H_0^c (\mathfrak{B} \mid U; \ \mathscr{F}_c) = \Gamma_c (U; \ \mathscr{F}).$$

Proof. Let $\mathfrak{U} = (U_i)$ be a locally finite covering of the set U by open sets belonging to the basis $\mathfrak{B} \mid U$. Let $f = (f_{i_0 \ldots i_k}) \in C_k^c(\mathfrak{U}; \mathscr{F}_c)$ be a cycle of dimension $k \geq 1$, i.e., $\partial f = 0$. By definition of a fine sheaf (cf. point 1.7.5) there exists a family of homomorphisms $e_i \colon \mathscr{F} \to \mathscr{F}$ with supports $\operatorname{Supp} e_i \subset U_i$ and with sum Σe_i, which is the identity isomorphism of the sheaf $\mathscr{F}$ onto itself. We set $g(g_{i_0 \ldots i_{k+1}}) \in C_{k+1}^c(\mathfrak{U}; \mathscr{F}_c)$, where

$$g_{i_0 i_1 \ldots i_{k+1}} = e_{i_0} f_{i_1 \ldots i_{k+1}}.$$

Then we get

$$(\partial g)_{i_0 \ldots i_k} = \sum_{j} e_j f_{i_0 \ldots i_k} + \sum_{j} \sum_{s=1}^{k+1} (-1)^s e_{i_0} f_{i_1 \ldots i_{s-1} j i_s \ldots i_k} = f_{i_0 \ldots i_k}.$$

Thus it is proved that

$$H_k^c(\mathfrak{U};\ \mathscr{F}_c) = 0 \quad \text{for} \quad k \neq 0.$$

Analogously one proves the existence of a natural isomorphism $H_0{}^c(\mathfrak{U};\ \mathscr{F}_c) = \Gamma_c(U;\ \mathscr{F})$. Now we note that there is a natural isomorphism of chain complexes

$$C_k^c(\mathfrak{B}\,|\,U;\ \mathscr{F}_c) = \varinjlim C_k^c(\mathfrak{U};\ \mathscr{F}_c),$$

where the inductive limit is taken with respect to the filtered set, ordered by inclusion, of all locally finite coverings $\mathfrak{U}$ of the set U by open sets belonging to the basis $\mathfrak{B}\,|\,U$. From this we get a natural isomorphism for homology groups

$$H_k^c(\mathfrak{B}\,|\,U;\ \mathscr{F}_c) = \varinjlim H_k^c(\mathfrak{U};\ \mathscr{F}_c).$$

The lemma is proved.

5.3. Let U_0 be an open set in X and E_{U_0} be a copresheaf on the category of all open sets in X, for which there exists an Abelian group G, such that $E_{U_0}(U) = G$ for $U_0 \subset U$ and $E_{U_0}(U) = 0$ for $U_0 \subset U$. We shall call the copresheaf E_{U_0} an *elementary copresheaf*.

Lemma. For any open sets U_0 and U in X,

$$H_k^c(\mathfrak{B}\,|\,U;\ E_{U_0}) = 0 \quad \text{for} \quad k \neq 0.$$

Proof. Let $\mathfrak{B}\,|\,U = (U_i)$ and let $I = \{i: U_0 \subset U_i\}$. Then

$$C_k^c(\mathfrak{B}\,|\,U;\ E_{U_0}) = \prod_{i_0, \ldots, i_k} G_{i_0 \ldots i_k},$$

where $G_{i_0 \ldots i_k} = G$, and the sum is taken over collections $(i_0, \ldots, i_k) \in I^{k+1}$. It follows from this that the complex $C.{}^c(\mathfrak{B}\,|\,U; E_{U_0})$ is acyclic in dimensions $k \geq 1$ (cf., e.g., Godement [1, p. 70]). The lemma is proved.

5.4. Let F be a copresheaf on the basis $\mathfrak{B}$. For each $U \in \mathfrak{B}$ we define an elementary copresheaf E_U for which $E_U(V) = F(U)$ for $U \subset V$ and $E_U(V) = 0$ for $U \subset V$. We denote by F_0 the copresheaf on the category of all open sets, which is the direct sum of the copresheaves $E_U(U \in \mathfrak{B})$. On the basis $\mathfrak{B}$ the natural homomorphisms $E_U \to F$ define a homomorphism $F_0 \to F$, which is, as is obvious directly, an epimorphism of copresheaves.

Taking the kernel of the epimorphism $F_0 \to F$ in place of F, and then continuing the process analogously, we get a resolution of the copresheaf F on the basis $\mathfrak{B}$:

$$\ldots \to F_1 \to F_0 \to F \to 0,$$

in which the copresheaves $F_0, F_1, \ldots$ are direct sums of elementary copresheaves. We shall call this resolution the *canonical resolution* of the copresheaf F.

Lemma. Suppose given an arbitrary resolution $F.$ of the copresheaf F on the basis $\mathfrak{B}$. Then for any open set $U \subset X$ there exists a spectral sequence

$$E^2_{p,\,q} = H_p H^c_q (\mathfrak{B} \,|\, U;\; F.) \Rightarrow H^c_{p+q} (\mathfrak{B} \,|\, U;\; F).$$

Proof. We consider the double complex

$$K_{p,\,q} = C^c_p (\mathfrak{B} \,|\, U;\; F_q).$$

We calculate the initial terms of both spectral sequences of this double complex. For the first spectral sequence

$$'E^1_{p,\,q} = 0 \quad \text{for} \quad q \neq 0,$$
$$'E^1_{p,\,0} = C^c_p (\mathfrak{B} \,|\, U;\; F).$$

Consequently, the first spectral sequence degenerates:

$$'E^2_{p,\,q} = 0 \quad \text{for} \quad q \neq 0;$$
$$'E^2_{p,\,0} = H^c_p (\mathfrak{B} \,|\, U;\; F).$$

For the second spectral sequence

$$''E^1_{p,\,q} = H^c_q (\mathfrak{B} \,|\, U;\; F_p);$$
$$''E^2_{p,\,q} = H_p H^c_q (\mathfrak{B} \,|\, U;\; F.).$$

The lemma is proved.

Corollary. Let $F.$ be the canonical resolution of the copresheaf F on the basis $\mathfrak{B}$. Then for any open set $U \subset X$ there is a natural isomorphism

$$H^c_k (\mathfrak{B} \,|\, U;\; F) = H_k H^c_0 (\mathfrak{B} \,|\, U;\; F.).$$

Proof. From Lemma 5.3 we get that under the assumptions made the spectral sequence of Lemma 5.4 degenerates. The assertion follows from this.

5.5. The copresheaf P on the basis $\mathfrak{B}$ is called *projective*, if for an arbitrary epimorphism $F_0 \to F$ each homomorphism $P \to F$ can be decomposed into the composition of a homomorphism $P \to F_0$ and the epimorphism $F_0 \to F$. In other words, the copresheaf P is projective, if each diagram with exact row

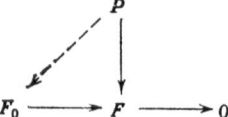

can be completed by the dashed arrow to a commutative diagram.

Lemma. If P is a projective copresheaf on the basis $\mathfrak{B}$, then for any open set $U \subset X$

$$H_k^c(\mathfrak{B} \mid U; \ P) = 0 \quad \text{for} \quad k \neq 0.$$

Proof. We consider the exact sequence of copresheaves on the basis $\mathfrak{B}$:

$$0 \to R \to P_0 \to P \to 0,$$

in which P_0 is the direct sum of elementary copresheaves, and R is the kernel of the epimorphism $P_0 \to P$. Since P is a projective copresheaf, this exact sequence splits. Consequently, the sequence

$$0 \to H_k^c(\mathfrak{B} \mid U; \ R) \to H_k^c(\mathfrak{B} \mid U; \ P_0) \to H_k^c(\mathfrak{B} \mid U; \ P) \to 0$$

is exact. On the other hand, by Lemma 5.3,

$$H_k^c(\mathfrak{B} \mid U; \ P_0) = 0 \quad \text{for} \quad k \neq 0.$$

The lemma is proved.

5.5.1. Let F be a copresheaf on the basis $\mathfrak{B}$. For each $U \in \mathfrak{B}$ we define an elementary copresheaf E_U, for which $E_U(V) = 0$ for $U \subset V$, and $E_U(V) = G$ for $U \subset V$, where G is the free Abelian group generated by the set $F(U)$. We denote by P_0 the copresheaf on $\mathfrak{B}$, which is the direct sum of the copresheaves $E_U (U \in \mathfrak{B})$. Then P_0 is a projective copresheaf and there is defined an epimorphism of copresheaves $P_0 \to F$ on the basis $\mathfrak{B}$. Continuing this process, we get a projective resolution of the copresheaf F[28]

$$\ldots \to P_1 \to P_0 \to F \to 0.$$

From Lemmas 5.4 and 5.5 we get that for any open set $U \subset X$ there is a natural isomorphism

$$H_k^c (\mathfrak{B} \mid U; \ F) = H_k H_0^c (\mathfrak{B} \mid U; \ P_\bullet),$$

i.e., the functor $F \mapsto H_k^c (\mathfrak{B} \mid U; F)$ is the left derived functor of the functor $F \to H_0^c (\mathfrak{B} \mid U; F)$.

5.6. Let F be an arbitrary copresheaf of Abelian groups on the basis $\mathfrak{B}$. For any open set $U \subset X$ we set

$$\check{F} (U) = H_0^c (\mathfrak{B} \mid U; \ F).$$

We get a copresheaf $\check{F}$ on the category of all open sets in X. If $U \in \mathfrak{B}$, then there is defined the natural homomorphism of Abelian groups $H_0^c (\mathfrak{B} \mid U; F) \to F(U)$, where for $U \subset U'$ the diagram

$$\begin{array}{ccc} H_0^c (\mathfrak{B} \mid U; \ F) & \to & F (U) \\ \downarrow & & \downarrow \\ H_0^c (\mathfrak{B} \mid U'; \ F) & \to & F (U') \end{array}$$

is commutative. Thus a homomorphism of copresheaves $\check{F} \to F$ on the basis $\mathfrak{B}$ is defined. Here, if $\mathscr{F}$ is a fine sheaf of Abelian groups on X, and $F = \mathscr{F}_c$, then by Lemma 5.2, $\check{F} = \mathscr{F}_c$ and the homomorphism $\check{F} \to F$ is the identity isomorphism of the copresheaf $\mathscr{F}_c$ onto itself.

5.7. Let F be a copresheaf of Abelian groups on the category of all open sets in the space X. We shall say that the copresheaf F is *regular*, if for each open set $U \subset X$ the natural map $\varinjlim F(V) \to F(U)$ is surjective, where the inductive limit is taken with respect to the filtered set, ordered by inclusion, of all relatively compact open sets $V \subset X$, whose closures are contained in U.

Lemma. For any copresheaf F on the basis $\mathfrak{B}$ there is a natural isomorphism

$$\lim_{\longrightarrow} H^c_k(\mathfrak{B}\,|\,V;\ F) = H^c_k(\mathfrak{B}\,|\,U;\ F),$$

where the inductive limit is taken with respect to the filtered set, ordered by inclusion, of all relatively compact open sets $V \subset X$, whose closures are contained in U.

Proof. Since $\mathfrak{B}\,|\,U$ consists of relatively compact open sets in X, whose closures are contained in U, there is a natural isomorphism of chain complexes

$$\lim_{\longrightarrow} C^c_k(\mathfrak{B}\,|\,V;\ F) = C^c_k(\mathfrak{B}\,|\,U;\ F),$$

where the inductive limit is taken with respect to the filtered set, ordered by inclusion, of all relatively compact open sets $V \subset X$, whose closures are contained in U. The assertion of the lemma follows from this.

Corollary. Let F be an arbitrary copresheaf of Abelian groups on the basis $\mathfrak{B}$. Then the copresheaf $\check{F}$ on the category of all open sets in X is regular.

5.8. Let F be an arbitrary copresheaf of Abelian groups on the category of all open sets in X. For an arbitrary compact set $M \subset X$ we set

$$G(M) = F(X)/F(X \setminus M).$$

Since for $M \subset M'$ the homomorphism $F(X \setminus M') \to F(X \setminus M)$ induces a homomorphism $G(M') \to G(M)$, one has that $G \colon M \to G(M)$ is a contravariant functor from the category of all compact sets of the space X into the category of Abelian groups. In other words, G is a presheaf of Abelian groups on the category of all compact sets of the space X. We shall call G *the presheaf associated with the copresheaf F.* Thus one can define an additive covariant functor $F \to G$ from the category of copresheaves to the category of presheaves. Obviously this functor is exact.

Lemma. Let $\mathscr{F}$ be a soft sheaf of Abelian groups on the topological space X. Then for the presheaf G, associated with the copresheaf $F = \mathscr{F}_c$, there is a natural isomorphism

$$G(M) = \Gamma(M;\ \mathscr{F}),$$

for any compact set $M \subset X$.

Proof. Since $\mathscr{F}$ is a soft sheaf, for any compact set $M \subset X$ there is an exact sequence of groups of sections

$$0 \longrightarrow \Gamma_c(X \setminus M; \mathscr{F}) \longrightarrow \Gamma_c(X; \mathscr{F}) \longrightarrow \Gamma(M; \mathscr{F}) \longrightarrow 0.$$

The assertion of the lemma follows directly from this.

5.9. Let F be a copresheaf of Abelian groups on the category of all open sets in X. Let G be a presheaf of Abelian groups on the category of all compact sets in X, associated with the copresheaf F. We denote by $\mathscr{G}$ the sheaf of Abelian groups on X generated by the presheaf G (cf. point 1.2.1). Then for each sheaf $x \in X$

$$\mathscr{G}_x = \lim_{\longrightarrow} G(M),$$

where the inductive limit is taken with respect to the filtered set, ordered by inclusion, of all compact neighborhoods M of the point x. By definition $\mathscr{G}$ is the union of the stalks $\mathscr{G}_x (x \in X)$ in which a topology is introduced naturally. We shall call $\mathscr{G}$ *the sheaf associated with the copresheaf* F. It follows from Lemma 5.8 that for a soft sheaf $\mathscr{F}$ the sheaf $\mathscr{G}$ associated with the copresheaf $F = \mathscr{F}_c$ coincides with $\mathscr{F}$. [29]

5.10. Lemma. Let F be a regular copresheaf on the category of all open sets in the space X. Let $\mathscr{G}$ be the sheaf of Abelian groups on X, associated with the copresheaf F. Then there exists a natural homomorphism $F \to \mathscr{G}_c$ of copresheaves on the category of all open sets in X. Here, if $\mathscr{F}$ is a soft sheaf of Abelian groups on X and $F = \mathscr{F}_c$, then $\mathscr{G} = \mathscr{F}$ and the homomorphism $F \to \mathscr{G}_c$ is the identity isomorphism of the copresheaf $\mathscr{F}_c$ onto itself.

Proof. Let U be an arbitrary open set in X and let f be an arbitrary element of the group $F(U)$. For an arbitrary point $x \in U$ we take a compact neighborhood $M \subset U$ and we denote by g_M the element of the group $G(M) = F(X)/F(X \setminus M)$, which is the image of the element f under the composite map $F(U) \to F(X) \to G(M)$. Let g_x be the element of the group $\mathscr{G}_x = \lim_{\longrightarrow} G(M)$, which is the image of the element g_M under the natural map $G(M) \to \mathscr{G}_x$. Obviously the element g_x is independent of the choice of neighborhood M of the point x. Obviously also the elements $g_x (x \in U)$ form a section $g: x \mapsto g_x$ of the sheaf $\mathscr{G}$ over the open set U. Since F is a regular copresheaf, there exists a relatively compact open set $U_0 \subset X$ such that $\bar{U}_0 \subset U$, and the element $f \in F(U)$ is the image of an element $f_0 \in F(U_0)$ under the map $F(U_0) \to F(U)$. Let M be a compact set contained in U, for which $M \cap U_0 = \varnothing$. Then $U_0 \subset X \setminus M$ and, consequently, there is a commutative diagram

$$
\begin{array}{ccc}
F(U_0) & \to & F(X \setminus M) \\
\downarrow & & \downarrow \\
F(U) & \to & F(X)
\end{array}
$$

It is directly evident from this diagram that the image of the element f in $G(M)$ is equal to zero: $g_M = 0$. Thus $g_x = 0$ for $x \notin \bar{U}_0$, i.e., the section $g \in \Gamma(U; \mathcal{G})$ has compact support contained in $\bar{U}_0$. Obviously the correspondence $f \mapsto g$ is a homomorphism of Abelian groups $F(U) \to \Gamma_c(U; \mathcal{G})$, which defines a homomorphism of copresheaves $F \to \mathcal{G}_c$. The lemma is proved.

5.11. Let F and G be copresheaves of Abelian groups on the basis $\mathfrak{B}$. We shall denote by $\mathrm{Hom}(\mathfrak{B}; F, G)$ the Abelian group of all homomorphisms $F \to G$.

Lemma. Let $\mathcal{F}$ be a soft sheaf of Abelian groups on the topological space X. Then for any sheaf of Abelian groups $\mathcal{G}$ on X there is a natural isomorphism

$$
\mathrm{Hom}(\mathfrak{B}; \mathcal{F}_c, \mathcal{G}_c) = \mathrm{Hom}(X; \mathcal{F}, \mathcal{G}).
$$

Proof. Since each homomorphism of sheaves of Abelian groups $\mathcal{F} \to \mathcal{G}$ defines a homomorphism $\mathcal{F}_c \to \mathcal{G}_c$ of copresheaves of Abelian groups on the basis $\mathfrak{B}$ in an obvious way, we get a natural homomorphism of Abelian groups

$$
\mathrm{Hom}(X; \mathcal{F}, \mathcal{G}) \to \mathrm{Hom}(\mathfrak{B}; \mathcal{F}_c, \mathcal{G}_c).
$$

Since $\mathcal{F}$ is a soft sheaf, this homomorphism is a monomorphism. It remains to be proved that it is an epimorphism. Suppose given a homomorphism $\mathcal{F}_c \to \mathcal{G}_c$ of copresheaves of Abelian groups on $\mathfrak{B}$. For an arbitrary point $x \in X$ we choose an open set $U \in \mathfrak{B}$ containing x. Then for each element $f_x \in \mathcal{F}_x$ there exists a section $f \in \Gamma_c(U; \mathcal{F})$, whose germ at the point x coincides with f_x. Let g be the image of the element f under the homomorphism $\Gamma_c(U; \mathcal{F}) \to \Gamma_c(U; \mathcal{G})$, defined by the homomorphism $\mathcal{F}_c$

$\rightarrow \mathscr{G}_c$. We denote by g_x the germ of the section g at the point x. We show that g_x is independent of the choice of open set U and section f. Let $f \in \Gamma_c(U; \mathscr{F})$ be a section over an arbitrary neighborhood $U \in \mathfrak{B}$ of the point $x \in X$, and let g be the image of the section f under the homomorphism $\Gamma_c(U; \mathscr{F}) \rightarrow \Gamma_c(U; \mathscr{G})$. It suffices to prove that $g_x = 0$ if $f_x = 0$. There exists a finite covering of the support $\operatorname{Supp} f$ of the section $f \in \Gamma_c(U; \mathscr{F})$ by open sets $U_i \in \mathfrak{B}$, contained in $U \setminus \{x\}$. Since $\mathscr{F}$ is a soft sheaf, there exists a decomposition $f = \Sigma f_i$, where $f_i \in \Gamma_c(U_i; \mathscr{F})$ (cf. Godement [1, p. 179]). We denote by g_i the image of the section f_i under the homomorphism $\Gamma_c(U_i; \mathscr{F}) \rightarrow \Gamma_c(U_i; \mathscr{G})$. Then $g = \Sigma g_i$ and, consequently, $g_x = 0$. Thus it is established that the correspondence $f_x \mapsto g_x$ defines a homomorphism $\mathscr{F} \rightarrow \mathscr{G}$ of sheaves of Abelian groups. The lemma is proved.

5.12. We consider the right derived functors

$$\operatorname{Ext}^k(\mathfrak{B}; F, G) = R^k \operatorname{Hom}(\mathfrak{B}; F, G)$$

of the functor $F \mapsto \operatorname{Hom}(\mathfrak{B}; F, G)$. Suppose given a projective resolution

$$\ldots \rightarrow P_1 \rightarrow P_0 \rightarrow F \rightarrow 0$$

of the cosheaf of Abelian groups F on the basis $\mathfrak{B}$. Then by definition

$$\operatorname{Ext}^k(\mathfrak{B}; F, G) = H^k \operatorname{Hom}(\mathfrak{B}; P_\cdot, G)$$

(cf., e.g., Grothendieck [2, p. 42]).

The exact sequence of copresheaves of Abelian groups on the basis $\mathfrak{B}$,

$$0 \rightarrow G \rightarrow F_{k-1} \rightarrow \ldots \rightarrow F_0 \rightarrow F \rightarrow 0,$$

is called a *k-fold exact sequence*, starting with the copresheaf G and ending with the copresheaf F. The *congruence relation* between two k-fold exact sequences starting with the copresheaf G and ending with the copresheaf F is the smallest reflexive, symmetric, and transitive relation following from the existence of a morphism of exact sequences with identity isomorphisms at the ends, i.e., following from the existence of a commutative diagram of the form

$$
\begin{array}{ccccccccc}
0 & \rightarrow & G & \rightarrow & F_{k-1} & \rightarrow \ldots \rightarrow & F_0 & \rightarrow & F & \rightarrow & 0 \\
& & \| & & \downarrow & & \downarrow & & \| & & \\
0 & \rightarrow & G & \rightarrow & F'_{k-1} & \rightarrow \ldots \rightarrow & F'_0 & \rightarrow & F & \rightarrow & 0.
\end{array}
$$

Let $E^k(F, G)$ be the set of classes of pairwise congruent k-fold exact sequences starting with the copresheaf G and ending with the copresheaf F. Then in the set $E^k(F, G)$ one can define an addition naturally (the so-called "Baer addition"), with respect to which $E^k(F, G)$ is an Abelian group (cf. MacLane [1, p. 116]). The zero element in the group $E^1(F, G)$ is the class of all split exact sequences of the form

$$0 \longrightarrow G \longrightarrow F_0 \longrightarrow F \longrightarrow 0.$$

The zero element in the group $E^k(F, G)$ for $k \geq 2$ is the congruence class of the k-fold exact sequence

$$0 \longrightarrow G \xrightarrow{\text{id}} G \longrightarrow 0 \longrightarrow \ldots \longrightarrow 0 \longrightarrow F \xrightarrow{\text{id}} F \longrightarrow 0.$$

There is a natural isomorphism of Abelian groups

$$E^k(F, \ G) = \text{Ext}^k(\mathfrak{B}; \ F, \ G)$$

(cf. MacLane [1, p. 121]).

5.13. Theorem. Let $\mathscr{F}$ be a fine sheaf of Abelian groups on the topological space X. Then for any injective sheaf of Abelian groups $\mathscr{G}$ on X

$$\text{Ext}^k(\mathfrak{B}; \ \mathscr{F}_c, \ \mathscr{G}_c) = 0 \quad \text{for} \quad k \neq 0.$$

Proof. First we consider the case $k = 1$. Suppose given an exact sequence

$$0 \longrightarrow \mathscr{G}_c \longrightarrow F_0 \longrightarrow \mathscr{F}_c \longrightarrow 0,$$

where F_0 is a copresheaf of Abelian groups on the basis $\mathfrak{B}$. From Lemma 5.2 and the definition of the homomorphism $\check{F}_0 \to F_0$ it follows that on the basis $\mathfrak{B}$ there is defined a commutative diagram

$$
\begin{array}{ccccccccc}
0 & \longrightarrow & \mathscr{G}_c & \longrightarrow & \check{F}_0 & \longrightarrow & \mathscr{F}_c & \longrightarrow & 0 \\
 & & \| & & \downarrow & & \| & & \\
0 & \longrightarrow & \mathscr{G}_c & \longrightarrow & F_0 & \longrightarrow & \mathscr{F}_c & \longrightarrow & 0
\end{array}
$$

with exact rows. By the five lemma the homomorphism $\check{F}_0 \to F_0$ on the basis $\mathfrak{B}$ is an isomorphism. In other words, the copresheaf $\check{F}_0$ is an extension of the copresheaf F_0 to the category of all open sets in X. Let $\mathscr{F}_0$ be the sheaf of Abelian groups on X, associated with the copresheaf $\check{F}_0$. Since the functor defined by the correspondence $F_0 \to \mathscr{F}_0$ is exact, in view of Lemma 5.8 we get an exact sequence of sheaves

$$0 \to \mathscr{G} \to \mathscr{F}_0 \to \mathscr{F} \to 0.$$

This sequence splits since $\mathscr{G}$ is an injective sheaf. Consequently, the exact sequence

$$0 \to \mathscr{G}_c \to \mathscr{F}_{0c} \to \mathscr{F}_c \to 0$$

also splits. On the other hand, by Lemma 5.10 there exists a commutative diagram with exact rows

$$
\begin{array}{ccccccccc}
0 & \to & \mathscr{G}_c & \to & \check{F}_0 & \to & \mathscr{F}_c & \to & 0 \\
& & \| & & \downarrow & & \| & & \\
0 & \to & \mathscr{G}_c & \to & \mathscr{F}_{0c} & \to & \mathscr{F}_c & \to & 0
\end{array}
$$

It follows from this that the homomorphism of copresheaves of Abelian groups $F_0 \to \mathscr{F}_{0c}$ is an isomorphism. Then $F_0 = \mathscr{F}_{0c} \mid \mathfrak{B}$ and, consequently, the exact sequence

$$0 \to \mathscr{G}_c \to F_0 \to \mathscr{F}_c \to 0$$

is split. Thus it is proved that $\mathrm{Ext}^1(\mathfrak{B}; \mathscr{F}_c, \mathscr{G}_c) = 0$.

Now we consider the case $k \geq 2$. Suppose given a k-fold exact sequence

$$0 \to \mathscr{G}_c \to F_{k-1} \to \ldots \to F_0 \to \mathscr{F}_c \to 0,$$

where $F_0, \ldots, F_{k-1}$ are copresheaves of Abelian groups on the basis $\mathfrak{B}$. One can assume that $F_0, \ldots, F_{k-2}$ are projective copresheaves of Abelian groups on the basis $\mathfrak{B}$ (cf. MacLane [1, p. 122]). From Lemma 5.2 and the definition of the homomorphisms $\check{F}_i \to F_i$ we get a commutative diagram

$$
\begin{array}{ccccccccccc}
0 & \to & \mathscr{G}_c & \to & \check{F}_{k-1} & \to & \ldots & \to & \check{F}_0 & \to & \mathscr{F}_c & \to & 0 \\
& & \| & & \downarrow & & & & \downarrow & & \| & & \\
0 & \to & \mathscr{G}_c & \to & F_{k-1} & \to & \ldots & \to & F_0 & \to & \mathscr{F}_c & \to & 0
\end{array}
$$

Since $\mathscr{F}$ is a fine sheaf, and $F_0, \ldots, F_{k-2}$ are projective copresheaves, the upper row in this diagram is exact. Thus, the rows are congruent k-fold exact sequences. Let $\mathscr{F}_0, \ldots, \mathscr{F}_{k-1}$ be the sheaves of Abelian groups on X, associated with the copresheaves $\check{F}_0, \ldots, \check{F}_{k-1}$. Then the sequence of sheaves of Abelian groups

$$0 \to \mathscr{G} \to \mathscr{F}_{k-1} \to \ldots \to \mathscr{F}_0 \to \mathscr{F} \to 0$$

is exact. Since the sheaf $\mathscr{G}$ is injective, there is defined a commutative diagram

$$\begin{array}{ccccccccccc}
0 & \to & \mathscr{G} & \to & \mathscr{F}_{k-1} & \to & \ldots & \to & \mathscr{F}_0 & \to & \mathscr{F} & \to & 0 \\
 & & \| & \text{id} & \downarrow & & & & \downarrow & \text{id} & \| \\
0 & \to & \mathscr{G} & \dashrightarrow & \mathscr{G} & \longrightarrow & \ldots & \to & \mathscr{F} & \dashrightarrow & \mathscr{F} & \to & 0
\end{array}$$

where the lower row is trivial. From this and Lemma 5.10 we get a commutative diagram

$$\begin{array}{ccccccccccc}
0 & \to & \mathscr{G}_c & \to & \check{F}_{k-1} & \longrightarrow & \ldots & \to & \check{F}_0 & \to & \mathscr{F}_c & \to & 0 \\
 & & \| & & \downarrow & & & & \downarrow & & \| \\
0 & \to & \mathscr{G}_c & \to & \mathscr{F}_{(k-1)c} & \longrightarrow & \ldots & \to & \mathscr{F}_{0c} & \to & \mathscr{F}_c & \to & 0 \\
 & & \| & & \downarrow & & & & \downarrow & & \| \\
0 & \to & \mathscr{G}_c & \to & \mathscr{G}_c & \longrightarrow & \ldots & \to & \mathscr{F}_c & \to & \mathscr{F}_c & \to & 0
\end{array}$$

in which the first and third rows are exact. Consequently, the first and third rows are congruent, and since the third row is trivial, $\text{Ext}^k(\mathfrak{B}; \mathscr{F}_c, \mathscr{G}_c) = 0$. The theorem is proved.

5.14. Theorem. Let $\mathscr{F}$ be a fine sheaf of Abelian groups on the topological space X. Then for any sheaf of Abelian groups $\mathscr{G}$ on X there exists a spectral sequence

$$E_2^{p,\,q} = \text{Ext}^p(\mathfrak{B}; \mathscr{F}_c, \mathscr{H}_c^q(\mathscr{G})) \Rightarrow \text{Ext}^{p+q}(X; \mathscr{F}, \mathscr{G}),$$

where $\mathscr{H}_c^q(\mathscr{G})$ is the copresheaf $U \mapsto H_c^q(U; \mathscr{G})$ on the category of all open sets in X.

Proof. We choose a projective resolution

$$\ldots \to P_1 \to P_0 \to \mathscr{F}_c \to 0$$

of the copresheaf $\mathscr{F}_c$ in the category of copresheaves of Abelian groups and an injective resolution

$$0 \to \mathscr{G} \to \mathscr{L}^0 \to \mathscr{L}^1 \to \ldots$$

of the sheaf $\mathscr{G}$ in the category of sheaves of Abelian groups. We consider the double complex

$$K^{p,\,q} = \operatorname{Hom}(\mathfrak{B};\ P_p,\ \mathscr{L}_c^q).$$

For the first spectral sequence of this double complex

$$'E_1^{p,\,q} = \operatorname{Hom}(\mathfrak{B};\ P_p,\ \mathscr{H}_c^q(\mathscr{G})),$$
$$'E_2^{p,\,q} = \operatorname{Ext}^p(\mathfrak{B};\ \mathscr{F}_c,\ \mathscr{H}_c^q(\mathscr{G})).$$

By Theorem 5.13 for the second spectral sequence

$$''E_1^{p,\,q} = \operatorname{Ext}^q(\mathfrak{B};\ \mathscr{F}_c,\ \mathscr{L}_c^p) = 0 \quad \text{for} \quad q \neq 0.$$

By Lemma 5.11 we get an isomorphism

$$''E_1^{p,\,0} = \operatorname{Hom}(X;\ \mathscr{F},\ \mathscr{L}^p).$$

Consequently, the second spectral sequence degenerates:

$$''E_2^{p,\,q} = 0 \quad \text{for} \quad q \neq 0,$$
$$''E_2^{p,\,0} = \operatorname{Ext}^p(X;\ \mathscr{F},\ \mathscr{G}).$$

The theorem is proved.

5.14.1. Corollary. For any fine sheaves of Abelian groups $\mathscr{F}$ and $\mathscr{G}$ on the topological space X there is a natural isomorphism

$$\operatorname{Ext}^k(\mathfrak{B};\ \mathscr{F}_c,\ \mathscr{G}_c) = \operatorname{Ext}^k(X;\ \mathscr{F},\ \mathscr{G}).$$

Proof. Since $\mathscr{G}$ is a fine sheaf, one has $\mathscr{H}_c^q(\mathscr{G}) = 0$ for $q \neq 0$. Hence, in this case the spectral sequence of Theorem 5.14 degenerates. The assertion follows from this.

5.15. Lemma. Let $\mathfrak{G}$ be a sheaf of commutative rings with unit on the topological space X. Then for any $\mathfrak{G}$-modules $\mathscr{F}$ and $\mathscr{G}$ there is a natural isomorphism

$$\operatorname{Ext}_{\mathfrak{G}}^k(X;\ \mathscr{F} \otimes_Z \mathfrak{G},\ \mathscr{G}) = \operatorname{Ext}^k(X;\ \mathscr{F},\ \mathscr{G}).$$

Proof. We make use of the natural isomorphism

$$\operatorname{Hom}_{\mathfrak{G}}(X;\ \mathscr{F} \otimes_Z \mathfrak{G},\ \mathscr{G}) = \operatorname{Hom}(X;\ \mathscr{F},\ \mathscr{G}).$$

Let an arbitrary injective resolution of the sheaf $\mathcal{G}$

$$0 \to \mathcal{G} \to \mathcal{L}^0 \to \mathcal{L}^1 \to \dots$$

in the category of $\mathcal{O}$-modules be chosen. It is also an injective resolution of the sheaf $\mathcal{G}$ in the category of sheaves of Abelian groups. The assertion follows directly from this.

5.16. Theorem. Let $\mathcal{O}$ be the sheaf of germs of holomorphic functions in the domain G of the space $\mathbb{C}^n$. Then for any fine sheaf of modules $\mathcal{F}$ on the sheaf of rings $\mathcal{O}$ in the domain G

$$\operatorname{Ext}^k_{\mathcal{O}}(G; \mathcal{F}, \mathcal{O}) = 0 \quad \text{for} \quad k \neq n.$$

Proof. Since $\operatorname{inj dim}_{\mathcal{O}}\mathcal{O} = n$ (by Theorem 4.4), it suffices to prove that $\operatorname{Ext}_{\mathcal{O}}{}^k(G; \mathcal{F}, \mathcal{O}) = 0$ for $k < n$. First we show that

$$\operatorname{Ext}^k_{\mathcal{O}}(G; \mathcal{F} \otimes_{\mathbb{Z}}\mathcal{O}, \mathcal{O}) = 0 \quad \text{for} \quad k < n.$$

By Lemma 5.15 there is an isomorphism

$$\operatorname{Ext}^k_{\mathcal{O}}(G; \mathcal{F} \otimes_{\mathbb{Z}}\mathcal{O}, \mathcal{O}) = \operatorname{Ext}^k(G; \mathcal{F}, \mathcal{O}).$$

Let $\mathfrak{B}$ be a basis of open sets in G, consisting of holomorphically complete relatively compact open sets. By Theorem 5.14 there exists a spectral sequence

$$E_2^{p, q} = \operatorname{Ext}^p(\mathfrak{B}, \mathcal{F}_c, \mathcal{H}_c^q(\mathcal{O})) \Rightarrow \operatorname{Ext}^{p+q}(G; \mathcal{F}, \mathcal{O}),$$

where $\mathcal{H}_c^q(\mathcal{O})$ is a copresheaf $U \mapsto H_c^q(U; \mathcal{O})$ on the basis $\mathfrak{B}$. By Corollary 3.5.3

$$\mathcal{H}_c^q(\mathcal{O}) = 0 \quad \text{for} \quad q \neq n.$$

Hence the spectral sequence degenerates. From this we get our assertion. We prove the assertion of the theorem by induction on k. For $k = 0$ the assertion is true since

$$\operatorname{Hom}_{\mathcal{O}}(G; \mathcal{F}, \mathcal{O}) = 0$$

for a fine sheaf $\mathcal{F}$. Let us assume that the assertion of the theorem is true for a $k < n - 1$. We make use of the exact sequence

$$0 \to \mathcal{R} \to \mathcal{F} \otimes_{\mathbb{Z}}\mathcal{O} \to \mathcal{F} \to 0,$$

where $\mathscr{R}$ is the kernel of the natural homomorphism $\mathscr{F} \otimes_z \mathcal{O} \to \mathscr{F}$. We get an exact sequence for the functor Ext:

$$\ldots \to \mathrm{Ext}_{\mathcal{O}}^{k}(G;\ \mathscr{R},\ \mathcal{O}) \to \mathrm{Ext}_{\mathcal{O}}^{k+1}(G;\ \mathscr{F},\ \mathcal{O}) \to \mathrm{Ext}_{\mathcal{O}}^{k+1}(G;\ \mathscr{F} \otimes_z \mathcal{O},\ \mathcal{O}) \to \ldots$$

Since $\mathscr{R}$ is obviously a fine sheaf, by the inductive hypothesis $\mathrm{Ext}_{\mathcal{O}}{}^{k}(G; \mathscr{R},\ \mathcal{O}) = 0$. From this and from what was proved at the very beginning, $\mathrm{Ext}_{\mathcal{O}}{}^{k+1}(G;\ \mathscr{F},\ \mathcal{O}) = 0$. Theorem 5.16 is proved completely.[30]

Chapter 2

HOMOLOGY THEORY

1. Sheaves of Germs of Homology

1.1. Let M be an analytic set in the domain G of the space $\mathbf{C}^n$. The set M becomes a ringed space by endowing it with the structure sheaf of complex algebras (cf. points 1.11.1 and 1.11.4 of Chapter 1)

$$\mathcal{O}_M = \mathcal{O}_G / \mathcal{I} \,|\, M,$$

where $\mathcal{O}_G$ is the sheaf of germs of homomorphic functions on G, and $\mathcal{I}$ is a coherent subsheaf of ideals in $\mathcal{O}_G$, whose set of zeros coincides with M. By an analytic sheaf on M, as usual, we mean a sheaf which is an $\mathcal{O}_M$-module.

Lemma. Let M be an analytic set in the domain G of the space $\mathbf{C}^n$. Let $\mathcal{F}$ and $\mathcal{G}$ be analytic sheaves on M, and $\mathcal{H}$ be an analytic sheaf on G. Then if the sheaf $\mathcal{G}$ is a flat $\mathcal{O}_M$-module, then there exists a spectral sequence

$$E_2^{p,\,q} = \operatorname{Ext}^p_{\mathcal{O}_M}(M; \mathcal{F}, \mathcal{E}xt^q_{\mathcal{O}_G}(\mathcal{G}^G, \mathcal{H})) \Rightarrow \operatorname{Ext}^{p+q}_{\mathcal{O}_G}(G; (\mathcal{F} \otimes_{\mathcal{O}_M} \mathcal{G})^G, \mathcal{H}).$$

Proof. Let $\mathcal{L}^{\boldsymbol{\cdot}}$ be an injective resolution of the $\mathcal{O}_G$-module $\mathcal{H}$. We choose an injective resolution of the complex $\mathcal{H}om_{\mathcal{O}_G}(\mathcal{G}^G, \mathcal{L}^{\boldsymbol{\cdot}})$ over the sheaf of rings $\mathcal{O}_M$ in the sense of Cartan–Eilenberg:

$$0 \to \mathcal{H}om_{\mathcal{O}_G}(\mathcal{G}^G, \mathcal{L}^{\boldsymbol{\cdot}}) \to \mathcal{L}^{0,\,\boldsymbol{\cdot}} \to \mathcal{L}^{1,\,\boldsymbol{\cdot}} \to \ldots$$

63

(cf. Cartan and Eilenberg [1, Chapter XVII, point 1, p. 437]). We consider the double complex

$$K^{p,q} = \operatorname{Hom}_{\mathcal{O}_M}(M;\ \mathcal{F},\ \mathcal{L}^{p,q}).$$

For its first spectral sequence we get

$${}'E_1^{p,q} = \operatorname{Hom}_{\mathcal{O}_M}(M;\ \mathcal{F},\ H^q\mathcal{L}^{p,\bullet}),$$

and since for fixed q we have an injective resolution

$$0 \to \mathcal{E}xt^q_{\mathcal{O}_G}(\mathcal{G}^G, \mathcal{H}) \to H^q\mathcal{L}^{0,\bullet} \to H^q\mathcal{L}^{1,\bullet} \to \ldots$$

of the $\mathcal{O}_M$-module $\mathcal{E}xt_{\mathcal{O}_G}{}^q(\mathcal{G}^G, \mathcal{H})$, one has

$${}'E_2^{p,q} = \operatorname{Ext}^p_{\mathcal{O}_M}(M;\ \mathcal{F},\ \mathcal{E}xt^q_{\mathcal{O}_G}(\mathcal{G}^G, \mathcal{H})).$$

For the second spectral sequence of the double complex $(K^{p,q})$, we get

$${}''E_1^{p,q} = \operatorname{Ext}^q_{\mathcal{O}_M}(M;\ \mathcal{F},\ \mathcal{H}om_{\mathcal{O}_G}(\mathcal{G}^G, \mathcal{L}^p)).$$

We show that for an injective $\mathcal{O}_G$-module $\mathcal{L}$ the $\mathcal{O}_M$-module $\mathcal{H}om_{\mathcal{O}_G}(\mathcal{G}^G, \mathcal{L})$ is also injective. In fact, since the sheaf $\mathcal{G}$ is a flat $\mathcal{O}_M$-module, each exact sequence of $\mathcal{O}_M$-modules $0 \to \mathcal{F} \to \mathcal{F}'$ defines an exact sequence

$$0 \to \mathcal{F} \otimes_{\mathcal{O}_M}\mathcal{G} \to \mathcal{F}' \otimes_{\mathcal{O}_M}\mathcal{G}$$

and a commutative diagram

$$\operatorname{Hom}_{\mathcal{O}_G}(G;\ (\mathcal{F}' \otimes_{\mathcal{O}_M}\mathcal{G})^G, \mathcal{L}) \to$$
$$\operatorname{Hom}_{\mathcal{O}_M}(M;\ \mathcal{F}',\ \mathcal{H}om_{\mathcal{O}_G}(\mathcal{G}^G, \mathcal{L})) \to$$
$$\to \operatorname{Hom}_{\mathcal{O}_G}(G;\ (\mathcal{F} \otimes_{\mathcal{O}_M}\mathcal{G})^G, \mathcal{L}) \to 0$$
$$\to \operatorname{Hom}_{\mathcal{O}_M}(M;\ \mathcal{F},\ \mathcal{H}om_{\mathcal{O}_G}(\mathcal{G}^G, \mathcal{L})) \to 0$$

The first row of this diagram is exact, since the sheaf $\mathcal{L}$ is injective as an $\mathcal{O}_G$-module. Consequently, the second row is also exact, and this means that the sheaf $\mathcal{H}om_{\mathcal{O}_G}(\mathcal{G}^G, \mathcal{L})$ is injective as an $\mathcal{O}_M$-module. Thus,

$${}''E_1^{p,q} = 0 \quad \text{for} \quad q \neq 0;$$
$${}''E_1^{p,0} = \operatorname{Hom}_{\mathcal{O}_G}(G;\ (\mathcal{F} \otimes_{\mathcal{O}_M}\mathcal{G})^G, \mathcal{L}^p).$$

Hence the second spectral sequence degenerates:

$$"E_2^{p,\,q} = 0 \quad \text{for} \quad q \neq 0;$$
$$"E_2^{p,\,0} = \text{Ext}_{\mathcal{O}_G}^p (G; (\mathscr{F} \otimes_{\mathcal{O}_M} \mathscr{G})^G, \mathscr{H}).$$

The lemma is proved.

1.2. Theorem. Let the analytic set M be realized as a subset of the domains G and G', respectively, of the complex spaces $\mathbf{C}^n$ and $\mathbf{C}^{n'}$. Then each holomorphic map $\varphi: G \to G''$ which induces the identity isomorphism of the set M as a ringed space onto itself induces for each analytic sheaf $\mathscr{F}$ on M an isomorphism of analytic sheaves.

$$\mathscr{E}\text{xt}_{\mathcal{O}_G}^{n-k} (\mathscr{F}^G, \Omega_G^n) \to \mathscr{E}\text{xt}_{\mathcal{O}_{G'}}^{n'-k} (\mathscr{F}^{G'}, \Omega_{G'}^{n'}),$$

where the latter is independent of the choice of holomorphic map φ.

Proof. Let

$$0 \overset{!}{\to} \Omega_G^n \to \mathscr{L}^0 \to \mathscr{L}^1 \to \dots$$

be an injective resolution of the $\mathcal{O}_G$-module Ω_G^n, and let

$$0 \to \mathscr{Hom}_{\mathcal{O}_G} (\mathcal{O}_M^G, \mathscr{L}^\bullet) \to \mathscr{L}^{0,\,\bullet} \to \mathscr{L}^{1,\,\bullet} \to \dots$$

be an injective resolution in the sense of Cartan–Eilenberg of the complex $\mathscr{Hom}_{\mathcal{O}_G}(\mathcal{O}_M^G, \mathscr{L}^\bullet)$ over the sheaf of rings $\mathcal{O}_M$. We consider the double complex

$$K_{p,\,q} = \mathscr{Hom}_{\mathcal{O}_M} (\mathscr{F}, \mathscr{L}^{-p,\,n-q}).$$

Using the natural isomorphism

$$\mathscr{Hom}_{\mathcal{O}_M} (\mathscr{F}, \mathscr{Hom}_{\mathcal{O}_G} (\mathcal{O}_M^G, \mathscr{L}^\bullet)) = \mathscr{Hom}_{\mathcal{O}_G} (\mathscr{F}^G, \mathscr{L}^\bullet),$$

we get a spectral sequence on M

$$E_{p,\,q}^2 = \mathscr{E}\text{xt}_{\mathcal{O}_M}^{-p} (\mathscr{F}, \mathscr{E}\text{xt}_{\mathcal{O}_G}^{n-q} (\mathcal{O}_M^G, \Omega_G^n)) \Rightarrow \mathscr{E}\text{xt}_{\mathcal{O}_G}^{n-p-q} (\mathscr{F}^G, \Omega_G^n)$$

(cf. the proof of Lemma 1.1). Analogously, for the domain G' we define the double complex

$$'K_{p,\,q} = \mathscr{Hom}_{\mathcal{O}_M} (\mathscr{F}, '\mathscr{L}^{-p,\,n'-q})$$

and on M we get a spectral sequence

$$'E^2_{p,\,q} = \mathscr{E}\mathrm{xt}^{-p}_{\mathcal{O}_M}(\mathscr{F},\,\mathscr{E}\mathrm{xt}^{n'-q}_{\mathcal{O}_{G'}}(\mathcal{O}^{G'}_M,\,\Omega^{n'}_{G'})) \Rightarrow \mathscr{E}\mathrm{xt}^{n'-p-q}_{\mathcal{O}_{G'}}(\mathscr{F}^{G'},\,\Omega^{n'}_{G'}).$$

1.2.1. Let $\mathscr{E}_G{}^q$ (respectively $\mathscr{E}_{G'}{}^q$) be the sheaf of germs of C^∞ differential forms of double degree $(0, q)$ on G (respectively on G'). A holomorphic map $\varphi\colon G \to G'$, which induces the identity isomorphism of the analytic set M onto itself, defines for each open set $U' \subset G'$ a homomorphism

$$\mathscr{E}^q_{G'}(U') \dashrightarrow \mathscr{E}^q_G(\varphi^{-1}(U')), \tag{1}$$

which is compatible with the restriction homomorphisms. Let $\mathscr{I}$ be a subsheaf of ideals in $\mathcal{O}_G$, for which $\mathcal{O}_M = \mathcal{O}_G/\mathscr{I}\,|\,M$. From the exact sequence

$$0 \to \mathscr{I}\mathscr{E}^q_G \to \mathscr{E}^q_G \to \mathscr{E}^q_G/\mathscr{I}\mathscr{E}^q_G \to 0$$

we get for an arbitrary open set $U \subset G$, an exact sequence

$$0 \to \Gamma_c(U;\,\mathscr{I}\mathscr{E}^q_G) \to \Gamma_c(U;\,\mathscr{E}^q_G) \to \Gamma_c(U;\,\mathscr{E}^q_G/\mathscr{I}\mathscr{E}^q_G) \to 0.$$

Passing to the dual spaces, we get an exact sequence

$$0 \to \Gamma'_c(U;\,\mathscr{E}^q_G/\mathscr{I}\mathscr{E}^q_G) \to '\mathscr{D}^{n-q}_G(U) \to '\mathscr{D}^{n-q}_G(U)/\mathscr{I}'\mathscr{D}^{n-q}_G(U) \to 0,$$

where $'\mathscr{D}_G{}^k$ is the sheaf of germs of currents of double degree (n, k) on G, and $\mathscr{I}'\mathscr{D}_G{}^k = \mathscr{H}\mathrm{om}_{\mathcal{O}_G}(\mathcal{O}_G/\mathscr{I},\,'\mathscr{D}_G{}^k)$. Thus, there is a natural isomorphism

$$\Gamma'_c(U;\,\mathscr{E}^q_G/\mathscr{I}\mathscr{E}^q_G) = \mathscr{I}'\mathscr{D}^{n-q}_G(U). \tag{2}$$

With the help of the homomorphism (1) we get a homomorphism

$$\Gamma_c(U';\,\mathscr{E}^q_{G'}/\mathscr{I}'\mathscr{E}^q_{G'}) \to \Gamma_c(\varphi^{-1}(U');\,\mathscr{E}^q_G/\mathscr{I}\mathscr{E}^q_G),$$

since $(\mathscr{E}_G{}^q/\mathscr{I}\mathscr{E}_G{}^q)\,|\,G\setminus M = 0$, and for each compact set $K \subset U'$ the intersection $\varphi^{-1}(K) \cap M = K \cap M$ is a compact set in $\varphi^{-1}(U')$. From the isomorphism (2) we get a homomorphism

$$\mathscr{I}'\mathscr{D}^{n-q}_G(\varphi^{-1}(U')) \to \mathscr{I}'\mathscr{D}^{n'-q}_{G'}(U'),$$

which is compatible with the restriction homomorphism. Since for the $\mathcal{O}_M$-modules $\mathscr{I}'\mathscr{D}_G{}^k$ and $\mathscr{I}'\mathscr{D}_{G'}{}^k$ we have, respectively, $\mathscr{I}'\mathscr{D}_G{}^k\,|\,G\setminus M = 0$ and $\mathscr{I}'\mathscr{D}_{G'}{}^k\,|\,G'\setminus M = 0$, we get a homomorphism of sheaves on M

$$\,_{g}'\mathcal{D}_{G}^{n-q} \to \,_{g'}'\mathcal{D}_{G'}^{n'-q}.$$

or, what is the same, a homomorphism

$$\mathcal{H}\mathrm{om}_{\mathcal{O}_{G}}(\mathcal{O}_{G}/\mathcal{I}, \,'\mathcal{D}_{G}^{n-q}) \to \mathcal{H}\mathrm{om}_{\mathcal{O}_{G'}}(\mathcal{O}_{G'}/\mathcal{I}', \,'\mathcal{D}_{G'}^{n'-q}). \qquad (3)$$

Let $\mathcal{L}^{\cdot}$ be an injective resolution of the complex $'\mathcal{D}^{\cdot}{}_{G}$, i.e., a quasi-isomorphism $'\mathcal{D}^{\cdot}{}_{G} \to \mathcal{L}^{\cdot}$ is defined. Then by Lemma 2.6 of Chapter 1, the complex $\mathcal{H}\mathrm{om}_{\mathcal{O}G}(\mathcal{O}_{G}/\mathcal{I}, \mathcal{L}^{\cdot})$ is an injective resolution of the complex $\mathcal{H}\mathrm{om}_{\mathcal{O}G}(\mathcal{O}_{G}/\mathcal{I}, \,'\mathcal{D}^{\cdot}{}_{G})$. The homomorphisms (3) define a homomorphism of complexes

$$\mathcal{H}\mathrm{om}_{\mathcal{O}_{G}}(\mathcal{O}_{G}/\mathcal{I}, \, \mathcal{L}^{n-q}) \to \mathcal{H}\mathrm{om}_{\mathcal{O}_{G'}}(\mathcal{O}_{G'}/\mathcal{I}', \, '\mathcal{L}^{n'-q}).$$

From this we get homomorphisms

$$\mathcal{L}^{-p,\,n-q} \longrightarrow \,'\mathcal{L}^{-p,\,n'-q}$$

and, consequently, a homomorphism of double complexes

$$K_{p,\,q} \longrightarrow \,'K_{p,\,q}. \qquad (4)$$

1.2.2. For each open set $U \subset G$ there is a natural isomorphism

$$H_{c}^{q}(U; \, \mathcal{O}_{G}/\mathcal{I}) = H_{c}^{q}(M \cap U; \, \mathcal{O}_{M}).$$

Consequently, a holomorphic map $\varphi: G \to G'$, which induces the identity isomorphism of the analytic set M onto itself, defines for each open set $U' \subset G'$ a natural homomorphism

$$H_{c}^{q}(U'; \, \mathcal{O}_{G'}/\mathcal{I}') \to H_{c}^{q}(\varphi^{-1}(U'); \, \mathcal{O}_{G}/\mathcal{I}),$$

which is actually defined by the identity isomorphism of the topological vector space $H_{c}{}^{q}(M \cap U'; \, \mathcal{O}_{M})$ onto itself. If U' is a holomorphically complete open set in G', then by Theorem 3.7 of Chapter 1, the dual map defines a natural isomorphism of vector spaces

$$\mathrm{Ext}_{\mathcal{O}_{G}}^{n-q}(\varphi^{-1}(U'); \, \mathcal{O}_{G}/\mathcal{I}, \Omega_{G}^{n}) \to \mathrm{Ext}_{\mathcal{O}_{G'}}^{n'-q}(U'; \, \mathcal{O}_{G'}/\mathcal{I}', \Omega_{G'}^{n'}).$$

Since this isomorphism is compatible with the restriction homomorphisms, passing to sheaves, we get a natural isomorphism of $\mathcal{O}_{M}$-modules

$$\mathscr{E}\mathrm{xt}_{6_G}^{n-q}\left(6_G/\mathfrak{I},\ \Omega_G^n\right)\longrightarrow\mathscr{E}\mathrm{xt}_{6_{G'}}^{n'-q}\left(6_{G'}/\mathfrak{I}',\ \Omega_{G'}^{n'}\right).$$

Thus, the homomorphism

$$E_{p,\,q}^2 \dashrightarrow {}'E_{p,\,q}^2$$

defined by the homomorphism (4) of double complexes is an isomorphism. Thus we get a natural isomorphism of analytic sheaves on M

$$\mathscr{E}\mathrm{xt}_{6_G}^{n-k}\left(\mathscr{F}^G,\ \Omega_G^n\right)\longrightarrow\mathscr{E}\mathrm{xt}_{6_{G'}}^{n'-k}\left(\mathscr{F}^{G'},\ \Omega_{G'}^{n'}\right).$$

The theorem is proved.

1.3. Let X be an arbitrary complex space and 6_X be its structure sheaf of complex algebras (cf. point 1.11.2 of Chapter 1). We choose a sufficiently fine covering (M_i) of the space X by connected open sets. Then for each i one can realize the set M_i as an analytic set in a domain G_i of the space $\mathbf{C}^{n_i}$. Here the restriction of the sheaf 6_X to M_i coincides with the sheaf 6_{M_i}. For each pair of indices (i, j) there exists a holomorphic map $\varphi_{ij} \colon G_{ij} \to G_{ji}$ of an open neighborhood $G_{ij} \subset G_i$ of the set $M_{ij} = M_i \cap M_j$ into an open neighborhood $G_{ji} \subset G_j$ of this same set, inducing the identity isomorphism of the set M_{ij} as a ringed space onto itself.

Let $\mathscr{F}$ be an arbitrary analytic sheaf on X. Then by Theorem 1.2 for each pair of indices (i, j) the holomorphic map $\varphi_{ij} \colon G_{ij} \to G_{ji}$ induces a natural isomorphism of analytic sheaves on M_{ij}

$$\rho_{ij}\colon\ \mathscr{E}\mathrm{xt}_{6_{G_i}}^{n_i-k}\left(\mathscr{F}^{G_i},\Omega_{G_i}^{n_i}\right)\longrightarrow\mathscr{E}\mathrm{xt}_{6_{G_j}}^{n_j-k}\left(\mathscr{F}^{G_j},\ \Omega_{G_j}^{n_j}\right).$$

Obviously ρ_{ii} is the identity isomorphism and for each triple of indices (i, j, k) over $M_{ijk} = M_i \cap M_j \cap M_k$ one has $\rho_{ik} = \rho_{jk} \circ \rho_{ij}$. It follows from this (cf. Serre [3, p. 376]), that on X there exists an analytic sheaf $\mathscr{H}_k(\mathscr{F})$, which is unique up to isomorphism, for which on each set M_i there is an isomorphism

$$\zeta_i\colon\ \mathscr{E}\mathrm{xt}_{6_{G_i}}^{n_i-k}\left(\mathscr{F}^{G_i},\ \Omega_{G_i}^{n_i}\right)\longrightarrow\mathscr{H}_k(\mathscr{F}),$$

while for each pair of indices (i, j), on the intersection M_{ij} there is a commutative diagram

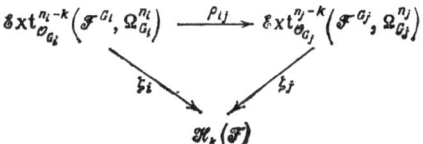

The analytic sheaf $\mathcal{H}_k(\mathcal{F})$ is called the *sheaf of germs of homology* of dimension k of the analytic sheaf $\mathcal{F}$ on the complex space X.[1]

1.3.1. For each exact sequence of analytic sheaves on the complex space X

$$0 \longrightarrow \mathcal{F}' \longrightarrow \mathcal{F} \longrightarrow \mathcal{F}'' \longrightarrow 0$$

there is defined a connecting homomorphism

$$\partial: \mathcal{H}_k(\mathcal{F}') \longrightarrow \mathcal{H}_{k-1}(\mathcal{F}'')$$

such that the sequence of analytic sheaves

$$\cdots \longrightarrow \mathcal{H}_k(\mathcal{F}'') \longrightarrow \mathcal{H}_k(\mathcal{F}) \longrightarrow \mathcal{H}_k(\mathcal{F}') \xrightarrow{\partial} \mathcal{H}_{k-1}(\mathcal{F}'') \longrightarrow \cdots$$

is exact.

In addition, if there is a commutative diagram with exact rows

$$\begin{array}{ccccccccc} 0 & \longrightarrow & \mathcal{F}' & \longrightarrow & \mathcal{F} & \longrightarrow & \mathcal{F}'' & \longrightarrow & 0 \\ & & \downarrow & & \downarrow & & \downarrow & & \\ 0 & \longrightarrow & \mathcal{G}' & \longrightarrow & \mathcal{G} & \longrightarrow & \mathcal{G}'' & \longrightarrow & 0 \end{array}$$

then the diagram

$$\begin{array}{ccc} \mathcal{H}_k(\mathcal{G}') & \xrightarrow{\partial} & \mathcal{H}_{k-1}(\mathcal{G}'') \\ \downarrow & & \downarrow \\ \mathcal{H}_k(\mathcal{F}') & \xrightarrow{\partial} & \mathcal{H}_{k-1}(\mathcal{F}'') \end{array}$$

is commutative.

Thus, the functor

$$\mathcal{F} \longmapsto \mathcal{H}_k(\mathcal{F})$$

is a contravariant homology functor in the sense of Grothendieck from the category of analytic sheaves on X to the category of analytic sheaves on X (cf. Grothendieck [2, p. 37]).

1.3.2. If G is a domain in the space $\mathbf{C}^n$, then by Theorem 4.4 of Chapter 1 the injective dimension of the $\mathcal{O}_G$-module $\Omega_G{}^n$ is equal to n. From this we get the following property of sheaves of germs of homology: *for any analytic sheaf $\mathcal{F}$ on the complex space X*

$$\mathcal{H}_k(\mathcal{F}) = 0 \quad \text{for} \quad k < 0.$$

1.3.3. If the analytic sheaf $\mathcal{F}$ on the complex space X is coherent, then the analytic sheaves $\mathcal{H}_k(\mathcal{F})$ ($k = 0, 1, \ldots$) are also coherent.[2]

1.3.4. An analytic sheaf $\mathcal{F}$ on a complex space X is said to be *locally fine*, if for each point of the space X there exists an open neighborhood, in which the analytic sheaf $\mathcal{F}$ is fine. From Theorem 5.16 of Chapter 1 we get: *if $\mathcal{F}$ is a locally fine analytic sheaf on the complex space X, then*

$$\mathcal{H}_k(\mathcal{F}) = 0 \quad \text{for} \quad k \neq 0.$$

2. Homology Groups

2.1. Lemma. For each locally fine analytic sheaf $\mathcal{F}$ on the complex space X, the sheaf $\mathcal{H}_0(\mathcal{F})$ is flabby.

Proof. Since the assertion of the lemma has local character, one can assume that $X = M$ is an analytic set in the domain G of the space $\mathbf{C}^n$. Let U be an open set in M. We make use of the exact sequence

$$\Gamma(M; \mathcal{H}_0(\mathcal{F})) \rightarrow \Gamma(U; \mathcal{H}_0(\mathcal{F})) \rightarrow H^1_S(M; \mathcal{H}_0(\mathcal{F})),$$

where $S = M \setminus U$. We show that $H_S{}^1(M; \mathcal{H}_0(\mathcal{F})) = 0$. In fact, there is a spectral sequence

$$E^{p,\,q}_2 = H^p_S(M; \mathcal{E}xt^q_{\mathcal{O}_G}(\mathcal{F}^G, \Omega^n_G)) \Rightarrow \mathrm{Ext}^{p+q}_{\mathcal{O}_G,\,S}(G; \mathcal{F}^G, \Omega^n_G)$$

(cf. point 1.10.2 of Chapter 1). Since $\mathcal{F}$ is a locally fine analytic sheaf, by Theorem 5.16 of Chapter 1

$$\mathcal{E}xt^q_{\mathcal{O}_G}(\mathcal{F}^G, \Omega^n_G) = 0 \quad \text{for} \quad q \neq n,$$

i.e., the spectral sequence degenerates. We get an isomorphism

$$H^p_\xi (M; \mathcal{H}_0(\mathcal{F})) = \text{Ext}^{p+n}_{\mathcal{O}_G \, s}(G; \mathcal{F}^G, \Omega^n_G).$$

On the other hand, the injective dimension of the $\mathcal{O}_G$-module Ω_{G^n} is equal to n (Theorem 4.4 of Chapter 1), so

$$\text{Ext}^{p+n}_{\mathcal{O}_G, \, s}(G; \mathcal{F}^G, \Omega^n_G) = 0 \quad \text{for} \quad p > 0$$

for any analytic sheaf $\mathcal{F}$ on M. The lemma is proved.

2.2. Let X be an arbitrary complex space and Φ be a family of supports in X (cf. point 1.7 of Chapter 1). For an arbitrary analytic sheaf $\mathcal{F}$ on X we set

$$H^\Phi_k (X; \ \mathcal{F}) = H_k \Gamma_\Phi (X; \ \mathcal{H}_0(\mathcal{C}^\cdot(\mathcal{F}))),$$

where $\mathcal{C}^\cdot(\mathcal{F})$ is the canonical resolution of Godement of the sheaf $\mathcal{F}$ (cf. point 1.7.3 of Chapter 1). The complex vector space $H_K{}^\Phi(X; \ \mathcal{F})$ is called the *homology group* of dimension k of the complex space X with coefficients in the sheaf $\mathcal{F}$ and with supports in the family Φ.[3]

Theorem. For each exact sequence of analytic sheaves on the complex space X

$$0 \longrightarrow \mathcal{F}' \longrightarrow \mathcal{F} \longrightarrow \mathcal{F}'' \longrightarrow 0$$

there is defined a connecting homomorphism

$$\partial \colon H^\Phi_k (X; \ \mathcal{F}') \longrightarrow H^\Phi_{k-1} (X; \ \mathcal{F}'')$$

such that the sequence of homology groups

$$\begin{aligned}
\ldots &\longrightarrow H^\Phi_k (X; \mathcal{F}'') \longrightarrow H^\Phi_k (X; \mathcal{F}) \longrightarrow H^\Phi_k (X; \mathcal{F}') \overset{\partial}{\longrightarrow} \\
&\overset{\partial}{\longrightarrow} H^\Phi_{k-1} (X; \mathcal{F}'') \longrightarrow \ldots \longrightarrow H^\Phi_0 (X; \mathcal{F}'') \longrightarrow \\
&\longrightarrow H^\Phi_0 (X; \mathcal{F}) \longrightarrow H^\Phi_0 (X; \mathcal{F}') \longrightarrow 0
\end{aligned}$$

is exact. Moreover, if there is a commutative diagram with exact rows

$$0 \to \mathscr{F}' \to \mathscr{F} \to \mathscr{F}'' \to 0$$
$$\downarrow \qquad \downarrow \qquad \downarrow$$
$$0 \to \mathscr{G}' \to \mathscr{G} \to \mathscr{G}'' \to 0$$

then the diagram

$$H_k^\Phi(X; \mathscr{G}') \xrightarrow{\partial} H_{k-1}^\Phi(X; \mathscr{G}'')$$
$$\downarrow \qquad\qquad \downarrow$$
$$H_k^\Phi(X; \mathscr{F}') \xrightarrow{\partial} H_{k-1}^\Phi(X; \mathscr{F}'')$$

is exact.

Proof. It suffices to prove the exactness of the sequence of chain complexes

$$0 \to \Gamma_\Phi(X; \mathscr{H}_0(\mathscr{C}^\cdot(\mathscr{F}''))) \to \Gamma_\Phi(X; \mathscr{H}_0(\mathscr{C}^\cdot(\mathscr{F}))) \to \Gamma_\Phi(X; \mathscr{H}_0(\mathscr{C}^\cdot(\mathscr{F}'))) \to 0.$$

The sequence of sheaves

$$0 \to \mathscr{C}^k(\mathscr{F}') \to \mathscr{C}^k(\mathscr{F}) \to \mathscr{C}^k(\mathscr{F}'') \to 0$$

is exact by definition, while the sheaf $\mathscr{C}^k(\mathscr{F}')$ is obviously locally fine. Consequently, the sequence

$$0 \to \mathscr{H}_0(\mathscr{C}^k(\mathscr{F}'')) \to \mathscr{H}_0(\mathscr{C}^k(\mathscr{F})) \to \mathscr{H}_0(\mathscr{C}^k(\mathscr{F}')) \to 0$$

is exact. By Lemma 2.1 the sheaf $\mathscr{H}_0(\mathscr{C}^k(\mathscr{F}''))$ is flabby, so the sequence

$$0 \to \Gamma_\Phi(X; \mathscr{H}_0(\mathscr{C}^k(\mathscr{F}''))) \to \Gamma_\Phi(X; \mathscr{H}_0(\mathscr{C}^k(\mathscr{F}))) \to \Gamma_\Phi(X; \mathscr{H}_0(\mathscr{C}^k(\mathscr{F}'))) \to 0$$

is exact. The theorem is proved.

Thus, the functor

$$\mathscr{F} \to H_k^\Phi(X; \mathscr{F})$$

is a contravariant homology functor in the sense of Grothendieck (cf. Grothendieck [2, p. 37]) from the category of analytic sheaves on X to the category of complex vector spaces.

2.3. If the family of supports Φ consists of all closed sets of the space X, then instead of $H_k^\Phi(X; \mathscr{F})$ we shall write $H_k(X; \mathscr{F})$.

Proposition. For each analytic sheaf $\mathcal{F}$ on the complex space X there is a real isomorphism

$$\mathcal{H}_k(\mathcal{F}) = H_k \mathcal{H}_0(\mathcal{C}^\cdot(\mathcal{F})).$$

In other words, the sheaf $\mathcal{H}_k(\mathcal{F})$ is naturally isomorphic to the sheaf generated by the presheaf $U \mapsto H_k(U; \mathcal{F})$.[4]

Proof. Since the assertion has local character, one can assume that $X = M$ is an analytic set in the domain G of the space $\mathbf{C}^n$. Since the injective dimension of the $\mathcal{O}_G$-module Ω_G^n is equal to n, the sheaf Ω_G^n has an injective resolution of length n:

$$0 \to \Omega_G^n \to \mathscr{L}^0 \to \mathscr{L}^1 \to \ldots \to \mathscr{L}^n \to 0.$$

We consider the double complex

$$K_{p,q} = \mathcal{H}\mathrm{om}_{\mathcal{O}_G}(\mathcal{C}^p(\mathcal{F})^G, \mathscr{L}^{n-q}).$$

For its first spectral sequence

$${}'E_{p,q}^1 = \mathcal{E}\mathrm{xt}_{\mathcal{O}_G}^{n-q}(\mathcal{C}^p(\mathcal{F})^G, \Omega_G^n)$$

and, consequently, ${}'E_{p,q}^1 = 0$ for $q \neq 0$, i.e., the first spectral sequence degenerates:

$${}'E_{p,q}^2 = 0 \quad \text{for} \quad q \neq 0, \qquad {}'E_{p,0}^2 = H_p \mathcal{H}_0(\mathcal{C}^\cdot(\mathcal{F})).$$

For the second spectral sequence

$${}''E_{p,q}^1 = 0 \quad \text{for} \quad q \neq 0, \qquad {}''E_{p,0}^1 = \mathcal{H}\mathrm{om}_{\mathcal{O}_G}(\mathcal{F}^G, \mathscr{L}^{n-p}),$$

i.e., the second spectral sequence also degenerates:

$${}''E_{p,q}^2 = 0 \quad \text{for} \quad q \neq 0, \quad {}''E_{p,0}^2 = \mathcal{H}_p(\mathcal{F}).$$

Thus the proposition is proved.

2.4. Theorem. If $\mathcal{F}$ is a locally fine analytic sheaf on the complex space X, then[5]

$$H_k^\Phi(X; \mathcal{F}) = 0 \quad \text{for} \quad k \neq 0;$$
$$H_0^\Phi(X; \mathcal{F}) = \Gamma_\Phi(X; \mathcal{H}_0(\mathcal{F})).$$

Proof. The analytic sheaf $\mathscr{F}$ has the property that the sheaf $\mathscr{H}om_{\mathcal{O}X}(\mathscr{F}, \mathscr{F})$ is locally soft. Since the sheaf $\mathscr{C}^0(\mathscr{F})$ is a $\mathscr{H}om_{\mathcal{O}X}(\mathscr{F}, \mathscr{F})$-module, the sheaf $\mathscr{X}^1 = \mathscr{C}^0(\mathscr{F})/\mathscr{F}$, i.e., $\mathscr{X}^1$ is a locally fine sheaf. From the exact sequence

$$0 \longrightarrow \mathscr{F} \longrightarrow \mathscr{C}^0(\mathscr{F}) \longrightarrow \mathscr{Z}^1 \longrightarrow 0$$

we get the exact sequence

$$0 \longrightarrow \mathscr{H}_0(\mathscr{Z}^1) \longrightarrow \mathscr{H}_0(\mathscr{C}^0(\mathscr{F})) \longrightarrow \mathscr{H}_0(\mathscr{F}) \longrightarrow 0.$$

Since the sheaf $\mathscr{H}_0(\mathscr{X}^1)$ is flabby, the sequence

$$0 \longrightarrow \Gamma_\Phi(X; \mathscr{H}_0(\mathscr{Z}^1)) \longrightarrow \Gamma_\Phi(X; \mathscr{H}_0(\mathscr{C}^0(\mathscr{F}))) \longrightarrow \Gamma_\Phi(X; \mathscr{H}_0(\mathscr{F})) \longrightarrow 0$$

is exact. Analogously, by induction, considering the exact sequence

$$0 \longrightarrow \mathscr{Z}^k \longrightarrow \mathscr{C}^k(\mathscr{F}) \longrightarrow \mathscr{Z}^{k+1} \longrightarrow 0,$$

where $\mathscr{C}^k(\mathscr{F}) = \mathscr{C}^0(\mathscr{X}^k)$ and $\mathscr{X}^{k+1} = \mathscr{C}^k(\mathscr{F})/\mathscr{X}^k$, we get the exact sequence

$$0 \longrightarrow \Gamma_\Phi(X; \mathscr{H}_0(\mathscr{Z}^{k+1})) \longrightarrow \Gamma_\Phi(X; \mathscr{H}_0(\mathscr{C}^k(\mathscr{F}))) \longrightarrow \Gamma_\Phi(X; \mathscr{H}_0(\mathscr{Z}^k)) \longrightarrow 0.$$

Thus the chain complex $\Gamma_\Phi(X; \mathscr{H}_0(\mathscr{C}^{\cdot}(\mathscr{F})))$ is acyclic. The theorem is proved.

2.5. Suppose given a complex of analytic sheaves on the complex space X

$$\cdots \longrightarrow \mathscr{L}^{q-1} \longrightarrow \mathscr{L}^q \longrightarrow \mathscr{L}^{q+1} \longrightarrow \cdots.$$

We consider the double complex

$$K_{p,q} = \Gamma_\Phi(X; \mathscr{H}_0(\mathscr{C}^p(\mathscr{L}^q))).$$

Since the functor

$$\mathscr{F} \longmapsto \Gamma_\Phi(X; \mathscr{H}_0(\mathscr{C}^p(\mathscr{F})))$$

is exact, for the first spectral sequence of the double complex $(K_{p,q})$ we get

$$'E^1_{p,q} = \Gamma_\Phi(X; \mathscr{H}_0(\mathscr{C}^p(H^q\mathscr{L}^{\cdot}))).$$

Consequently,

$$'E_{p,q}^2 = H_p^\Phi (X; \; H^q \mathscr{L}^\bullet).$$

Obviously for the second spectral sequence of the complex $(K_{p,q})$

$$''E_{p,q}^1 = H_q^\Phi (X; \; \mathscr{L}^p);$$
$$''E_{p,q}^2 = H_p H_q^\Phi (X; \; \mathscr{L}^\bullet).$$

Suppose given, for example, a locally fine resolution

$$0 \to \mathscr{F} \to \mathscr{L}^0 \to \mathscr{L}^1 \to \dots$$

of the analytic sheaf $\mathscr{F}$ on X. Then the first spectral sequence degenerates:

$$'E_{p,q}^2 = 0 \quad \text{for} \quad q \neq 0, \quad 'E_{p,0}^2 = H_p^\Phi (X; \; \mathscr{F}).$$

The second spectral sequence also degenerates:

$$''E_{p,q}^2 = 0 \quad \text{for} \quad q \neq 0, \quad ''E_{p,0}^2 = H_p \Gamma_\Phi (X; \; \mathscr{H}_0(\mathscr{L}^\bullet)).$$

Thus we have proved the following

Proposition. For each locally fine resolution $\mathscr{L}^\bullet$ on the complex space X there is a natural isomorphism

$$H_k^\Phi (X; \; \mathscr{F}) = H_k \Gamma_\Phi (X; \; \mathscr{H}_0(\mathscr{L}^\bullet)).$$

2.6. Let X be an arbitrary topological space, U be an open set in X, $S = X \setminus U$. Then for any flabby sheaf $\mathscr{L}$ on X there is an exact sequence

$$0 \to \Gamma_{\Phi|S}(X; \; \mathscr{L}) \to \Gamma_\Phi (X; \; \mathscr{L}) \to \Gamma_{\Phi \cap U}(U; \; \mathscr{L}) \to 0.$$

Setting $\mathscr{L} = \mathscr{C}^k(\mathscr{F})$ here, we get an exact sequence of cochain complexes, from which we can derive the familiar exact sequence for cohomology groups, connected with the open set U (cf. point 1.8.2 of Chapter 1). If X is a complex space and $\mathscr{F}$ is an analytic sheaf on X, then setting $\mathscr{L} = \mathscr{H}_0(\mathscr{C}^k(\mathscr{F}))$, we get an exact sequence of chain complexes, from which we derive the following exact sequence for homology groups:

$$\dots \to H_{k+1}^{\Phi \cap U}(U; \; \mathscr{F}) \to H_k^{\Phi|S}(X; \; \mathscr{F}) \to H_k^\Phi (X; \; \mathscr{F}) \to H_k^{\Phi \cap U}(U; \; \mathscr{F}) \to \dots .$$

If the topological space X is paracompact or if Φ is a paracompactifying family of supports (cf. Godement [1, p. 172]), then for any flabby sheaf $\mathscr{L}$ on X there is an exact sequence

$$0 \longrightarrow \Gamma_{\Phi|U}(U; \; \mathscr{L}) \longrightarrow \Gamma_{\Phi}(X; \; \mathscr{L}) \longrightarrow \Gamma_{\Phi \cap S}(S; \; \mathscr{L}) \longrightarrow 0.$$

Setting $\mathscr{L} = \mathscr{C}^k(\mathscr{F})$ here, we derive the familiar exact sequence for cohomology groups connected with the closed set S (cf. point 1.8.1 of Chapter 1). If X is a complex space and $\mathscr{F}$ is an analytic sheaf on X, then under these assumptions setting $\mathscr{L} = \mathscr{H}_0(\mathscr{C}^k(\mathscr{F}))$, we derive the following exact sequence for homology groups:

$$\ldots \longrightarrow H_{k+1}^{\Phi \cap S}(S; \; \mathscr{F}) \longrightarrow H_k^{\Phi|U}(U; \; \mathscr{F}) \longrightarrow H_k^{\Phi}(X; \; \mathscr{F}) \longrightarrow H_k^{\Phi \cap S}(S; \; \mathscr{F}) \longrightarrow \ldots.$$

3. Connection with the Functors Ext

3.1. Theorem. Suppose given on the complex space X a locally fine resolution of the structure sheaf $\mathcal{O}_X$

$$0 \longrightarrow \mathcal{O}_X \longrightarrow \mathscr{L}^0 \dashrightarrow \mathscr{L}^1 \longrightarrow \ldots,$$

consisting of flat $\mathcal{O}_X$-modules. Then there is a natural isomorphism

$$H_k^{\Phi}(X; \; \mathscr{F}) = H_k \mathrm{Hom}_{\mathcal{O}_X, \, \Phi}(X; \; \mathscr{F}, \; \mathscr{H}_0(\mathscr{L}^{\cdot}))$$

for any analytic sheaf $\mathscr{F}$ on X.

In particular,

$$\mathscr{H}_k(\mathscr{F}) = H_k \mathscr{H}om_{\mathcal{O}_X}(\mathscr{F}, \; \mathscr{H}_0(\mathscr{L}^{\cdot})).$$

Proof. For an arbitrary analytic sheaf $\mathscr{F}$ on X, we set

$$\mathscr{L}_{\mathscr{F}}^k = \mathscr{F} \otimes_{\mathcal{O}_X} \mathscr{L}^k.$$

We get a sequence of locally fine analytic sheaves on X. Since $\mathscr{L}^k$ are flat $\mathcal{O}_X$-modules, we get a resolution of the sheaf $\mathscr{F}$

$$0 \dashrightarrow \mathscr{F} \longrightarrow \mathscr{L}_{\mathscr{F}}^0 \longrightarrow \mathscr{L}_{\mathscr{F}}^1 \longrightarrow \ldots,$$

consisting of locally fine analytic sheaves. By virtue of Proposition 2.5, we get a natural isomorphism

$$H_k^{\Phi}(X; \; \mathscr{F}) = H_k \Gamma_{\Phi}(X; \; \mathscr{H}_0(\mathscr{L}_{\mathscr{F}}^{\cdot})).$$

Now let a neighborhood of an arbitrary point of the space X be realized as an analytic set M in a domain G of the space $\mathbf{C}^n$. In this neighborhood there is a natural isomorphism

$$\mathcal{H}_0\left(\mathcal{L}_{\mathcal{F}}^k\right) = \mathcal{E}\mathrm{xt}_{\mathcal{O}_G}^n\left(\mathcal{L}_{\mathcal{F}}^{kG}, \ \Omega_G^n\right).$$

By virtue of Lemma 1.1 there exists a spectral sequence

$$E_2^{p,q} = \mathcal{E}\mathrm{xt}_{\mathcal{O}_M}^p\left(\mathcal{F}, \ \mathcal{E}\mathrm{xt}_{\mathcal{O}_G}^q\left(\mathcal{L}^{kG}, \ \Omega_G^n\right)\right) \Rightarrow \mathcal{E}\mathrm{xt}_{\mathcal{O}_G}^{p+q}\left(\mathcal{L}_{\mathcal{F}}^{kG}, \ \Omega_G^n\right).$$

Since $\mathcal{L}^k$ is a locally fine analytic sheaf, one has

$$\mathcal{E}\mathrm{xt}_{\mathcal{O}_G}^q\left(\mathcal{L}^{kG}, \ \Omega_G^n\right) = 0 \quad \text{for} \quad q \neq n.$$

Hence the spectral sequence degenerates, and one gets an isomorphism

$$\mathcal{E}\mathrm{xt}_{\mathcal{O}_M}^p\left(\mathcal{F}, \ \mathcal{E}\mathrm{xt}_{\mathcal{O}_G}^n\left(\mathcal{L}^{kG}, \ \Omega_G^n\right)\right) = \mathcal{E}\mathrm{xt}_{\mathcal{O}_G}^{p+n}\left(\mathcal{L}_{\mathcal{F}}^{kG}, \ \Omega_G^n\right),$$

which for $p = 0$ reduces to an isomorphism

$$\mathcal{H}\mathrm{om}_{\mathcal{O}_M}\left(\mathcal{F}, \ \mathcal{H}_0\left(\mathcal{L}^k\right)\right) = \mathcal{H}_0\left(\mathcal{L}_{\mathcal{F}}^k\right).$$

The theorem is proved.

Remark. As a locally fine resolution of the sheaf $\mathcal{O}_X$, consisting of flat $\mathcal{O}_X$-modules, one can take, for example, the canonical resolution of Godement $\mathcal{C}^{\cdot}(\mathcal{O}_X)$. One can also take an arbitrary locally fine resolution of the simple sheaf $\mathbf{C}$ on X:

$$0 \longrightarrow \mathbf{C} \longrightarrow \mathcal{S}^0 \longrightarrow \mathcal{S}^1 \longrightarrow \ldots.$$

Then the sheaves $\mathcal{L}^k = \mathcal{O}_X \otimes_{\mathbf{C}} \mathcal{S}^k$ are locally fine flat $\mathcal{O}_X$-modules. Consequently, the locally fine "Cartan resolution" $\mathcal{L}^{\cdot}$ of the sheaf $\mathcal{O}_X$ consists of flat $\mathcal{O}_X$-modules.

3.2. Lemma. If the locally fine analytic sheaf $\mathcal{L}$ on the complex space X is a flat $\mathcal{O}_X$-module, then the sheaf $\mathcal{H}_0(\mathcal{L})$ is an injective $\mathcal{O}_X$-module.

Proof. Since the injectivity of a sheaf is a local property, one can assume that $X = M$ is an analytic set in the domain G of the space $\mathbf{C}^n$. For an arbitrary analytic sheaf $\mathcal{F}$ on M, by Lemma 1.1 there exists a spectral sequence

$$E_i^{p,\,q} = \mathrm{Ext}_{\mathcal{O}_M}^p \left(M;\, \mathscr{F},\, \mathcal{E}\mathrm{xt}_{\mathcal{O}_G}^q (\mathscr{L}^G, \Omega_G^n) \right) \Rightarrow \mathrm{Ext}_{\mathcal{O}_G}^{p+q} (G;\, \mathscr{L}_{\mathscr{F}}^G, \Omega_G^n),$$

where $\mathscr{L}_{\mathscr{F}} = \mathscr{F} \otimes_{\mathcal{O}_M} \mathscr{L}$. Since

$$\mathcal{E}\mathrm{xt}_{\mathcal{O}_G}^q (\mathscr{L}^G,\, \Omega_G^n) = 0 \quad \text{for} \quad q \neq n,$$

we get an isomorphism

$$\mathrm{Ext}_{\mathcal{O}_M}^p \left(M;\, \mathscr{F},\, \mathcal{E}\mathrm{xt}_{\mathcal{O}_G}^n (\mathscr{L}^G,\, \Omega_G^n) \right) = \mathrm{Ext}_{\mathcal{O}_G}^{p+n} (G;\, \mathscr{L}_{\mathscr{F}}^G,\, \Omega_G^n).$$

Since the injective dimension of the $\mathcal{O}_G$-module Ω_G^n is equal to n, one has $\mathrm{Ext}_{\mathcal{O}_G}{}^k(G;\, \mathscr{L}_{\mathscr{F}}{}^G, \Omega_G{}^n) = 0$ for $k > n$, so

$$\mathrm{Ext}_{\mathcal{O}_M}^p \left(M;\, \mathscr{F},\, \mathscr{H}_0(\mathscr{L}) \right) = 0 \quad \text{for} \quad p \neq 0.$$

The lemma is proved.

3.3. Theorem. For any analytic sheaf $\mathscr{F}$ on the complex space X, there exists a spectral sequence

$$E_{p,\,q}^2 = \mathrm{Ext}_{\mathcal{O}_X,\,\Phi}^{-p} (X;\, \mathscr{F},\, \mathscr{H}_q(\mathcal{O}_X)) \Rightarrow H_{p+q}^\Phi (X;\, \mathscr{F}).$$

Proof. We choose a locally fine resolution

$$0 \to \mathcal{O}_X \to \mathscr{L}^0 \to \mathscr{L}^1 \to \cdots$$

of the sheaf $\mathcal{O}_X$, consisting of flat $\mathcal{O}_X$-modules. We construct an injective resolution in the sense of Cartan–Eilenberg

$$0 \to \mathscr{H}_0(\mathscr{L}^{\cdot}) \to \mathscr{L}_{0,\,\cdot} \to \mathscr{L}_{-1,\,\cdot} \to \cdots$$

of the complex $\mathscr{H}_0(\mathscr{L}^{\cdot})$ (cf. Cartan and Eilenberg [1, Chapter XVII, point 1, p. 437]). We consider the double complex

$$K_{p,\,q} = \mathrm{Hom}_{\mathcal{O}_X,\,\Phi}(X;\, \mathscr{F},\, \mathscr{L}_{p,\,q}),$$

where $\mathscr{F}$ is an arbitrary analytic sheaf on X. For its first spectral sequence

$${}'E_{p,\,q}^1 = \mathrm{Hom}_{\mathcal{O}_X,\,\Phi}(X;\, \mathscr{F},\, H_q \mathscr{L}_{p,\,\cdot}).$$

Since for fixed q we get an injective resolution

$$0 \to H_q \mathcal{H}_0(\mathcal{L}^{\cdot}) \to H_q \mathcal{L}_{0,\cdot} \to H_q \mathcal{L}_{-1,\cdot} \to \cdots$$

of the sheaf $H_q \mathcal{H}_0(\mathcal{L}^{\cdot}) = \mathcal{H}_q(\mathcal{O}_X)$ (cf. Proposition 2.5), one has

$${}'E_{p,q}^2 = \operatorname{Ext}_{\mathcal{O}_X, \Phi}^{-p}(X; \mathcal{F}, \mathcal{H}_q(\mathcal{O}_X)).$$

For the second spectral sequence of the double complex $(K_{p,q})$, one has

$${}''E_{p,q}^1 = \operatorname{Ext}_{\mathcal{O}_X, \Phi}^{-q}(X; \mathcal{F}, \mathcal{H}_0(\mathcal{L}^p)).$$

By Lemma 3.2 the $\mathcal{O}_X$-module $\mathcal{H}_0(\mathcal{L}^p)$ is injective, so

$${}''E_{p,q}^1 = 0 \quad \text{for} \quad q \neq 0.$$

Consequently, the second spectral sequence degenerates:

$${}''E_{p,q}^2 = 0 \quad \text{for} \quad q \neq 0;$$
$${}''E_{p,0}^2 = H_p \operatorname{Hom}_{\mathcal{O}_X, \Phi}(X; \mathcal{F}, \mathcal{H}_0(\mathcal{L}^{\cdot})).$$

The assertion now follows from Theorem 3.1.

3.4. Proposition. Let X be a complex manifold of dimension n. Then for any analytic sheaf $\mathcal{F}$ on X there is a natural isomorphism

$$H_k^\Phi(X; \mathcal{F}) = \operatorname{Ext}_{\mathcal{O}_X, \Phi}^{n-k}(X; \mathcal{F}, \Omega_X^n).$$

In particular,

$$\mathcal{H}_k(\mathcal{F}) = \mathscr{E}\mathrm{xt}_{\mathcal{O}_X}^{n-k}(\mathcal{F}, \Omega_X^n).$$

Proof. We consider the spectral sequence of Theorem 3.3. Since X is a complex manifold, one has

$$\mathcal{H}_q(\mathcal{O}_X) = 0 \quad \text{for} \quad q \neq n$$

(cf. Corollary 3.5.3 of Chapter 1). Consequently, the spectral sequence degenerates. The assertion of Proposition 3.4 follows directly from this.

Corollary. If M is an analytic set in the domain G of the space $\mathbf{C}^n$, then for any analytic sheaf $\mathcal{F}$ on M there is a natural isomorphism

$$H_k^\Phi(M; \mathcal{F}) = \operatorname{Ext}_{\mathcal{O}_G, \Phi}^{n-k}(G; \mathcal{F}^G, \Omega_G^n).$$

Proof. By virtue of the exact sequence of point 2.6, there is an isomorphism

$$H_k^{\Phi}(M;\ \mathscr{F}) = H_k^{\Phi}(G;\ \mathscr{F}^G).$$

The assertion follows directly from this in view of Proposition 3.4.

4. Poincaré Duality

4.1. Theorem. For any analytic sheaf $\mathscr{F}$ on the complex space X there exists a spectral sequence[6]

$$E_{p,\,q}^2 = H_{\Phi}^{-p}(X;\ \mathscr{H}_q(\mathscr{F})) \Rightarrow H_{p+q}^{\Phi}(X;\ \mathscr{F}).$$

Proof. Let

$$0 \to \mathscr{F} \to \mathscr{L}^0 \to \mathscr{L}^1 \to \dots$$

be a locally fine resolution of the analytic sheaf $\mathscr{F}$. We choose an injective resolution in the sense of Cartan–Eilenberg

$$0 \to \mathscr{H}_0(\mathscr{L}^{\cdot}) \to \mathscr{L}_{0,\,.} \to \mathscr{L}_{-1,\,.} \to \dots$$

of the complex $\mathscr{H}_0(\mathscr{L}^{\cdot})$ and we consider the double complex

$$K_{p,\,q} = \Gamma_{\Phi}(X;\ \mathscr{L}_{p,\,q}).$$

For its first spectral sequence

$$'E_{p,\,q}^1 = \Gamma_{\Phi}(X;\ H_q\mathscr{L}_{p,\,.}).$$

Since for fixed q we have an injective resolution

$$0 \to H_q\mathscr{H}_0(\mathscr{L}^{\cdot}) \to H_q\mathscr{L}_{0,\,.} \to H_q\mathscr{L}_{-1,\,.} \to \dots$$

of the sheaf $H_q\mathscr{H}_0(\mathscr{L}^{\cdot}) = \mathscr{H}_q(\mathscr{F})$ (cf. Proposition 2.5), we have

$$'E_{p,\,q}^2 = H_{\Phi}^{-p}(X;\ \mathscr{H}_q(\mathscr{F})).$$

On the other hand, for the second spectral sequence

$$''E_{p,\,q}^1 = H_{\Phi}^{-q}(X;\ \mathscr{H}_0(\mathscr{L}^p)).$$

By Lemma 2.1 the sheaf $\mathscr{H}_0(\mathscr{L}^p)$ is flabby, so

$$''E^1_{p,q} = 0 \quad \text{for} \quad q \neq 0.$$

Thus the second spectral sequence degenerates:

$$''E^2_{p,q} = 0 \quad \text{for} \quad q \neq 0;$$
$$''E^2_{p,0} = H_p\Gamma_\Phi(X; \mathscr{H}_0(\mathscr{L}')) = H^\Phi_p(X; \mathscr{F}).$$

The theorem is proved.

4.2. Corollary (Poincaré duality). Let X be a complex manifold of dimension n and let $\mathscr{F}$ be a locally free analytic sheaf on X. Then there exists a natural isomorphism

$$H^\Phi_k(X; \mathscr{F}) = H^{n-k}_\Phi(X; \mathscr{H}_n(\mathscr{F})),$$

while[7]

$$\mathscr{H}_n(\mathscr{F}) = \mathscr{H}om_{\mathcal{O}_X}(\mathscr{F}, \Omega^n_X).$$

Proof. Since the space X is a complex manifold, and the sheaf $\mathscr{F}$ is locally free, one has

$$\mathscr{H}_q(\mathscr{F}) = 0 \quad \text{for} \quad q \neq n$$

(Corollary 3.5.3 of Chapter 1). In such a case the spectral sequence of Theorem 4.1 degenerates. The assertion is proved.

4.3. Proposition. Let $\mathscr{F}$ be an arbitrary analytic sheaf on the complex space X. Then Poincaré duality

$$H_k(U; \mathscr{F}) = H^{n-k}(U; \mathscr{H}_n(\mathscr{F}))$$

holds for any open set $U \subset X$ if and only if[8]

$$\mathscr{H}_k(\mathscr{F}) = 0 \quad \text{for} \quad k \neq n.$$

Proof. Let Poincaré duality hold for any open set in X. Then, passing to the inductive limit with respect to the filtered set of all open neighborhoods of an arbitrary point $x \in X$, we get a natural isomorphism

$$\mathscr{H}_k(\mathscr{F})_x = H^{n-k}(\{x\}; \mathscr{H}_n(\mathscr{F}))$$

(cf. Proposition 2.3, and also Godement [1, p. 219]). On the other hand, it is obvious that

$$H^k(\{x\}; \mathscr{H}_n(\mathscr{F})) = 0 \quad \text{for} \quad k \neq 0;$$
$$H^0(\{x\}; \mathscr{H}_n(\mathscr{F})) = \mathscr{H}_n(\mathscr{F})_x.$$

Thus the necessity of the condition is proved. The sufficiency follows directly from Theorem 4.1.

4.4. Lemma. Let X be a reduced complex space. Then in order that the point $x \in X$ be regular, it is necessary and sufficient that

$$\mathscr{H}_k(\mathscr{O}_X)_x = 0 \quad \text{for} \quad k \neq n,$$

where n is the dimension of the Zariski tangent space (cf., e.g., Hartshorne [2, p. 60]) of the space X at the point x.

Proof. Since the hypotheses and the assertion of the lemma have local character, one can assume that $X = M$ is an analytic set in the domain G of the space $\mathbf{C}^n$. If the point x is regular, then $M = G$, and

$$H_k(G; \mathscr{O}_G) = H^{n-k}(G; \Omega_G^n)$$

(cf. Corollary 4.2). The necessity of the condition follows from this. Conversely, if the condition of the lemma holds, then

$$\mathscr{H}_k(\mathscr{O}_M) = 0 \quad \text{for} \quad k < n$$

in a neighborhood of the point x. Let y be a regular point of the set M, lying in a neighborhood of the point x. Then by what has been proved, $\mathscr{H}_k(\mathscr{O}_M) = 0$ for $k \neq d$, where $d = \dim_y M$, while $\mathscr{H}_d(\mathscr{O}_M) \neq 0$ Since $d \leq n$, by the assumption made $d = n$. This means that the point x is regular. The lemma is proved.

4.5. From Proposition 4.3 with the help of Lemma 4.4 we get the following

Proposition. Let X be a reduced complex space. Then Poincaré duality

$$H_k(U; \mathscr{O}_X) = H^{n-k}(U; \mathscr{H}_n(\mathscr{O}_X))$$

holds for the structure sheaf $\mathscr{O}_X$ and any open set $U \subset X$ if and only if the space X is a complex manifold of dimension n.

5. Dualizing Complexes[9]

5.1. Suppose given on each complex space X a complex of $\mathcal{O}_X$-modules $\mathcal{K}^{\cdot}_X$ such that the following conditions hold:

a) if G is a domain in the space $\mathbf{C}^n$, then each stalk $\mathcal{K}^p_{G,z}$ is an injective $\mathcal{O}_{G,z}$-module ($z \in G$); moreover,

$$H^p \mathcal{K}^{\cdot}_G = 0 \quad \text{for} \quad p \neq -n,$$
$$H^{-n} \mathcal{K}^{\cdot}_G = \Omega^n_G,$$

where Ω_G^n is the sheaf of germs of holomorphic n-forms on G;

b) if M is an analytic set in the domain G of the space $\mathbf{C}^n$, then on M there exists a natural isomorphism[10]

$$\mathcal{K}^{\cdot}_M = \mathcal{H}om_{\mathcal{O}_G}(\mathcal{O}^G_M, \mathcal{K}^{\cdot}_G).$$

The complex $\mathcal{K}^{\cdot}_X$ is called the *dualizing complex* of the space X (cf. Ramis and Ruget [1]).[11]

5.2. Lemma. Let M be an analytic set in the domain G of the space $\mathbf{C}^n$ and $\mathcal{K}^{\cdot}_M$ be a dualizing complex of the set M. Then for any analytic sheaf $\mathcal{F}$ on M there is a natural isomorphism[12]

$$\mathcal{E}xt^{-p}_{\mathcal{O}_M}(\mathcal{F}, \mathcal{K}^{\cdot}_M) = \mathcal{H}_p(\mathcal{F}).$$

Proof. We make use of the spectral sequence

$$E^{p,q}_2 = \mathcal{E}xt^p_{\mathcal{O}_G}(\mathcal{F}, H^q \mathcal{K}^{\cdot}_G) \Rightarrow \mathcal{E}xt^{p+q}_{\mathcal{O}_G}(\mathcal{F}, \mathcal{K}^{\cdot}_G),$$

where $\mathcal{F}$ is an arbitrary $\mathcal{O}_G$-module (cf., e.g., Grothendieck [2, p. 47]). Since $H^q \mathcal{K}^{\cdot}_G = 0$ for $q \neq -n$, this spectral sequence degenerates. We get a natural isomorphism

$$\mathcal{E}xt^{-p}_{\mathcal{O}_G}(\mathcal{F}, \mathcal{K}^{\cdot}_G) = \mathcal{E}xt^{n-p}_{\mathcal{O}_G}(\mathcal{F}, \Omega^n_G).$$

In particular, setting $\mathcal{F} = \mathcal{O}_M^G$, we get an isomorphism

$$H^{-p} \mathcal{K}^{\cdot}_M = \mathcal{E}xt^{-p}_{\mathcal{O}_G}(\mathcal{O}^G_M, \mathcal{K}^{\cdot}_G) = \mathcal{H}_p(\mathcal{O}_M).$$

Let $\mathscr{L}^{\cdot}_G$ be an injective resolution of the complex $\mathscr{K}_G$. Then the injective complex of $\mathcal{O}_M$-modules

$$\mathscr{L}^{\cdot}_M = \mathscr{H}\text{om}_{\mathcal{O}_G}(\mathcal{O}^G_M, \mathscr{L}^{\cdot}_G)$$

is a resolution of the complex $\mathscr{K}^{\cdot}_M$. Moreover, there is a natural isomorphism

$$\mathscr{H}\text{om}_{\mathcal{O}_M}(\mathscr{F}, \mathscr{L}^{\cdot}_M) = \mathscr{H}\text{om}_{\mathcal{O}_M}(\mathscr{F}, \mathscr{H}\text{om}_{\mathcal{O}_G}(\mathcal{O}^G_M, \mathscr{L}^{\cdot}_G)) = \mathscr{H}\text{om}_{\mathcal{O}_G}(\mathscr{F}^G, \mathscr{L}^{\cdot}_G),$$

where $\mathscr{F}$ is an arbitrary analytic sheaf on M. Consequently, passing to cohomology, we get an isomorphism

$$\mathscr{E}\text{xt}^{-p}_{\mathcal{O}_M}(\mathscr{F}, \mathscr{K}^{\cdot}_M) = \mathscr{E}\text{xt}^{-p}_{\mathcal{O}_G}(\mathscr{F}^G, \mathscr{K}^{\cdot}_G).$$

Thus the lemma is proved.

5.3. Lemma. Let $\mathscr{K}^{\cdot}_X$ be a dualizing complex for the complex space X, Φ be a family of supports in X. Then for any locally fine $\mathcal{O}_X$-module $\mathscr{F}$ on X,

$$\text{Ext}^k_{\mathcal{O}_{X'}, \Phi}(X; \mathscr{F}, \mathscr{K}^{\cdot}_X) = 0 \quad \text{for} \quad k \neq 0;$$
$$\text{Ext}^0_{\mathcal{O}_{X'}, \Phi}(X; \mathscr{F}, \mathscr{K}^{\cdot}_X) = \Gamma_\Phi(X; \mathscr{H}_0(\mathscr{F})).$$

Proof. We make use of the spectral sequence

$$E^{p, q}_2 = H^p_\Phi(X; \mathscr{E}\text{xt}^q_{\mathcal{O}_X}(\mathscr{F}, \mathscr{K}^{\cdot}_X)) \Rightarrow \text{Ext}^{p+q}_{\mathcal{O}_X, \Phi}(X; \mathscr{F}, \mathscr{K}^{\cdot}_X)$$

(cf. Grothendieck [2, p. 114]). By Lemma 5.2 there is an isomorphism

$$\mathscr{E}\text{xt}^q_{\mathcal{O}_X}(\mathscr{F}, \mathscr{K}^{\cdot}_X) = \mathscr{H}_{-q}(\mathscr{F}).$$

Since $\mathscr{F}$ is a locally fine $\mathcal{O}_X$-module, $\mathscr{H}_q(\mathscr{F}) = 0$ for $q \neq 0$. Consequently, the spectral sequence degenerates. We get an isomorphism

$$\text{Ext}^k_{\mathcal{O}_{X}, \Phi}(X; \mathscr{F}, \mathscr{K}^{\cdot}_X) = H^k_\Phi(X; \mathscr{H}_0(\mathscr{F})).$$

By Lemma 2.1 the sheaf $\mathscr{H}_0(\mathscr{F})$ is flabby. The lemma is proved.

5.4. Theorem. Let $\mathscr{K}^{\cdot}_X$ be a dualizing complex of the complex space X, Φ be an arbitrary family of supports in X. Then for any analytic sheaf $\mathscr{F}$ on X there is a natural isomorphism

$$\text{Ext}_{\mathcal{O}_X, \Phi}^{-p}(X; \mathscr{F}, \mathscr{K}_X^{\cdot}) = H_p^{\Phi}(X; \mathscr{F}).$$

Proof. We consider the double complex

$$K_{p, q} = \text{Hom}_{\mathcal{O}_X, \Phi}(X; \mathscr{L}^p, \mathscr{L}_{\bar{X}}^q),$$

where $\mathscr{L}^{\cdot}$ is a locally fine resolution of the $\mathcal{O}_X$-module $\mathscr{F}$, and $\mathscr{L}^{\cdot}{}_X$ is an injective resolution of the dualizing complex $\mathscr{K}^{\cdot}{}_X$. For the first spectral sequence of this double complex we get

$$'E_{p, q}^1 = \text{Ext}_{\mathcal{O}_X, \Phi}^{-q}(X; \mathscr{L}^p, \mathscr{K}_X^{\cdot}).$$

By Lemma 5.3

$$'E_{p, q}^1 = 0 \quad \text{for} \quad q \neq 0,$$
$$'E_{p, 0}^1 = \Gamma_{\Phi}(X; \mathscr{H}_0(\mathscr{L}^p)).$$

Consequently, by Proposition 2.5,

$$'E_{p, q}^2 = 0 \quad \text{for} \quad q \neq 0,$$
$$'E_{p, 0}^2 = H_p^{\Phi}(X; \mathscr{F}).$$

For the second spectral sequence of the double complex $(K_{p,q})$ we get

$$''E_{p, q}^1 = 0 \quad \text{for} \quad q \neq 0,$$
$$''E_{p, 0}^1 = \text{Hom}_{\mathcal{O}_X, \Phi}(X; \mathscr{F}, \mathscr{L}_{\bar{X}}^p).$$

Consequently,

$$''E_{p, q}^2 = 0 \quad \text{for} \quad q \neq 0,$$
$$''E_{p, 0}^2 = \text{Ext}_{\mathcal{O}_X, \Phi}^{-p}(X; \mathscr{F}, \mathscr{K}_X^{\cdot}).$$

The theorem is proved.

Corollary. If the dualizing complex $\mathscr{K}^{\cdot}{}_X$ is injective, then for any analytic sheaf $\mathscr{F}$ on X there is a natural isomorphism

$$H_p^{\Phi}(X; \mathscr{F}) = H^{-p}\text{Hom}_{\mathcal{O}_X, \Phi}(X; \mathscr{F}, \mathscr{K}_X^{\cdot}).$$

5.5. Theorem. Suppose there is defined on each complex space X a locally fine resolution

$$0 \to \mathcal{O}_X \to \mathscr{L}_X^0 \to \mathscr{L}_X^1 \to \dots,$$

consisting of flat $\mathcal{O}_X$-modules, such that the following condition holds:

If M is an analytic set in the domain G of the space $\mathbf{C}^n$, then on M there exists a natural isomorphism

$$\mathcal{L}_M^{\cdot} = \mathcal{O}_M^q \otimes_{\mathcal{O}_G} \mathcal{L}_G^{\cdot}.$$

Then the $\mathcal{O}_X$-modules

$$\mathcal{K}_X^{-p} = \mathcal{H}_0(\mathcal{L}_X^p)$$

form an injective dualizing complex of the space X.[13]

Proof. The complex $\mathcal{K}_X^{\cdot}$ is injective by Lemma 3.2. On the other hand, by Proposition 2.5 there is an isomorphism

$$H^{-p}\mathcal{K}_X^{\cdot} = \mathcal{H}_p(\mathcal{O}_X).$$

In particular, if G is a domain in the space $\mathbf{C}^n$, then

$$H^{-p}\mathcal{K}_G^{\cdot} = \mathcal{E}xt_{\mathcal{O}_G}^{n-p}(\mathcal{O}_G, \Omega_G^n).$$

Let M be an analytic set in the domain G of the space $\mathbf{C}^n$. Then by Lemma 1.1 there exists a spectral sequence

$$E_2^{p,\,q} = \mathcal{E}xt_{\mathcal{O}_G}^p(\mathcal{O}_M^q, \mathcal{E}xt_{\mathcal{O}_G}^q(\mathcal{L}_G^{\cdot}, \Omega_G^n)) \Rightarrow \mathcal{E}xt_{\mathcal{O}_G}^{p+q}(\mathcal{O}_M^q \otimes_{\mathcal{O}_G} \mathcal{L}_G^{\cdot}, \Omega_G^n).$$

Since

$$\mathcal{E}xt_{\mathcal{O}_G}^q(\mathcal{L}_G^{\cdot}, \Omega_G^n) = 0 \quad \text{for} \quad q \neq n,$$

this spectral sequence degenerates. We get an isomorphism

$$\mathcal{H}om_{\mathcal{O}_G}(\mathcal{O}_M^q, \mathcal{H}_0(\mathcal{L}_G^{\cdot})) = \mathcal{H}_0(\mathcal{L}_M).$$

The theorem is proved.

Remark. The hypotheses of Theorem 5.5 are satisfied, for example, by the canonical resolution of Godement $\mathcal{C}^{\cdot}(\mathcal{O}_X)$ of the structure sheaf $\mathcal{O}_X$.[14]

Chapter 3

DUALITY THEOREMS

1. Homology of Coverings

1.1. Let X be a locally compact topological space, countable at infinity. A covariant functor F, defined on the category of all open sets in X, and assuming values in the category of Abelian groups, is called a *copresheaf* of Abelian groups on the category of open sets in X. For each locally finite covering $\mathfrak{U} = (U_i)$ of the space X with respect to compact open sets one can define the *group of chains* with compact supports of the covering $\mathfrak{U}$ with coefficients in the copresheaf F

$$C_k^c(\mathfrak{U}; F) = \prod_{i_0, \ldots, i_k} F(U_{i_0 \ldots i_k}),$$

where the direct sum is taken over those collections of indices $(i_0, \ldots, i_k)$ for which the intersection $U_{i_0 \ldots i_k} = U_{i_0} \cap \ldots \cap U_{i_k}$ is not empty, the *boundary operator*

$$\partial \colon C_k^c(\mathfrak{U}; F) \to C_{k-1}^c(\mathfrak{U}; F),$$

and the *homology groups* with compact supports of the covering $\mathfrak{U}$ with coefficients in the copresheaf F:

$$H_k^c(\mathfrak{U}; F) = H_k C_{\bullet}^c(\mathfrak{U}; F)$$

(cf. point 5.1 of Chapter 1).

Theorem. Let X be a complex space, countable at infinity, and $\mathfrak{U} = (U_i)$ be a locally finite covering of the space X by relatively compact open sets. Then for any analytic sheaf $\mathcal{F}$ on X there exists a spectral sequence

$$E^2_{p,\,q} = H^c_p\,(\mathfrak{U};\,\mathcal{H}^c_q\,(\mathcal{F})) \Rightarrow H^c_{p+q}\,(X;\,\mathcal{F}),$$

where $\mathcal{H}_q{}^c(\mathcal{F})$ is the copresheaf $U \mapsto H_q{}^c(U;\,\mathcal{F})$ on the category of open sets in X.[1]

Proof. For an arbitrary analytic sheaf $\mathcal{L}$ on X we set

$$\mathcal{C}_p\,(\mathcal{L}) = \prod_{i_0,\,\ldots,\,i_p} \mathcal{L}_{U_{i_0}\ldots i_p},$$

where the product is taken over those collections of indices $(i_0, \ldots, i_p)$ for which the intersection $U_{i_0\ldots i_p} = U_{i_0} \cap \ldots \cap U_{i_p}$ is not empty. If $l = (l_{i_0\ldots i_p}) \in \mathcal{C}_p(\mathcal{L})$, then let

$$(\partial l)_{i_0\,\ldots\,i_{p-1}} = \sum_j \sum_{k=0}^p (-1)^k\, l_{i_0\,\ldots\,i_{k-1}ji_k\,\ldots\,i_{p-1}},$$

where j runs through those values for which the intersection $U_j \cap U_{i_0\ldots i_{p-1}}$ is not empty. Thus there is defined a homomorphism of sheaves

$$\partial\colon\ \mathcal{C}_p\,(\mathcal{L}) \to \mathcal{C}_{p-1}(\mathcal{L}),$$

where the sequence

$$\ldots \xrightarrow{\partial} \mathcal{C}_1\,(\mathcal{L}) \xrightarrow{\partial} \mathcal{C}_0\,(\mathcal{L}) \to \mathcal{L} \to 0$$

is exact. We consider the double complex

$$K_{p,\,q} = \Gamma_c\,(X;\,\mathcal{C}_p\,(\mathcal{H}_0\,(\mathcal{L}^q))) = \prod_{i_0,\,\ldots,\,i_p} \Gamma_c\,(U_{i_0\,\ldots\,i_p};\ \mathcal{H}_0\,(\mathcal{L}^q)),$$

where

$$0 \to \mathcal{F} \to \mathcal{L}^0 \to \mathcal{L}^1 \to \ldots$$

is a locally fine resolution of the analytic sheaf $\mathcal{F}$. For the first spectral sequence of the double complex $(K_{p,q})$, by virtue of Proposition 2.5 of Chapter 2 we get

$$'E^1_{p,\,q} = \prod_{i_0,\,\ldots,\,i_p} H^c_q(U_{i_0\,\ldots\,i_p};\,\mathscr{F}).$$

Consequently,

$$'E^2_{p,\,q} = H^c_p(\mathfrak{U};\,\mathscr{H}^c_q(\mathscr{F})),$$

where $\mathscr{H}_q^c(\mathscr{F})$ is the copresheaf $U \mapsto H_q^c(U;\,\mathscr{F})$ on the category of open sets in X. On the other hand, for the second spectral sequence of the double complex $(K_{p,q})$

$$"E^1_{p,\,q} = 0 \quad \text{for} \quad q \neq 0;$$
$$"E^1_{p,\,0} = \Gamma_c(X;\,\mathscr{H}_0(\mathscr{L}^p)),$$

since the sheaves $\mathscr{H}_0(\mathscr{L}^p)$ are flabby (cf. Lemma 2.1 of Chapter 2). Hence the second spectral sequence degenerates:

$$"E^2_{p,\,q} = 0 \quad \text{for} \quad q \neq 0;$$
$$"E^2_{p,\,0} = H^c_p(X;\,\mathscr{F}).$$

The theorem is proved.

1.2. Let X be a locally compact topological space, countable at infinity. A covariant functor F, defined on the category of all closed sets in X and assuming values in the category of Abelian groups, is called a *copresheaf* of Abelian groups on the category of closed sets in X. For each covering of finite type $\mathfrak{M} = (M_i)$ of the space X by closed sets one can define the *group of chains* of the covering $\mathfrak{M}$ with coefficients in the copresheaf F

$$C_k(\mathfrak{M};\,F) = \prod_{i_0,\,\ldots,\,i_k} F(M_{i_0\,\ldots\,i_k}),$$

where the product is taken over those collections of indices $(i_0, \ldots, i_k)$ for which the intersection $M_{i_0\ldots i_k} = M_{i_0} \cap \ldots \cap M_{i_k}$ is not empty, the *boundary operator*

$$\partial\colon C_k(\mathfrak{M};\,F) \to C_{k-1}(\mathfrak{M};\,F),$$

and the *homology groups* of the covering $\mathfrak{M}$ with coefficients in the copresheaf F:

$$H_k(\mathfrak{M};\,F) = H_k C_\bullet(\mathfrak{M};\,F)$$

(cf. point 5.1 of Chapter 1).

Theorem. Let X be a complex space, countable at infinity, and $\mathfrak{M} = (M_i)$ be a locally finite covering of the space X by compact sets. Then for any analytic sheaf $\mathscr{F}$ on X there exists a spectral sequence

$$E^2_{p,\,q} = H_p\left(\mathfrak{M};\ \mathscr{H}^X_q(\mathscr{F})\right) \Rightarrow H_{p+q}(X;\ \mathscr{F}),$$

where $\mathscr{H}_q^X(\mathscr{F})$ is the copresheaf $M \mapsto H_q M(X;\ \mathscr{F})$ on the category of closed sets in X.

Proof. Let $\mathscr{L}$ be an arbitrary analytic sheaf on X. For an arbitrary closed set $M \subset X$ we consider the sheaf $_M\mathscr{L}$ of germs of sections of the sheaf $\mathscr{L}$ with supports in M. Then

$$\Gamma(U;\ _M\mathscr{L}) = \Gamma_M(U;\ \mathscr{L})$$

for any open set $U \subset X$. We set

$$\mathscr{C}_p(\mathscr{L}) = \prod_{i_0,\,\ldots,\,i_p}\ _{M_{i_0}\ldots i_p}\mathscr{L},$$

where the product is taken over those collections of indices $(i_0, \ldots, i_p)$ for which the intersection $M_{i_0\ldots i_p} = M_{i_0} \cap \ldots \cap M_{i_p}$ is not empty. In the usual way we define a homomorphism of sheaves

$$\partial\colon\ \mathscr{C}_p(\mathscr{L}) \longrightarrow \mathscr{C}_{p-1}(\mathscr{L}),$$

namely: if $l = (l_{i_0\ldots i_p}) \in \mathscr{C}_p(\mathscr{L})$, then

$$(\partial l)_{i_0\ldots i_{p-1}} = \sum_j \sum_{k=0}^p (-1)^k l_{i_0\ldots i_{k-1} j i_k \ldots i_{p-1}},$$

where the index j runs through those values for which the intersection $M_j \cap M_{i_0\ldots i_{p-1}}$ is not empty. It is easy to prove that the sequence of sheaves

$$\ldots \xrightarrow{\partial} \mathscr{C}_1(\mathscr{L}) \xrightarrow{\partial} \mathscr{C}_0(\mathscr{L}) \longrightarrow \mathscr{L} \longrightarrow 0$$

is exact. We consider the double complex

$$K_{p,\,q} = \Gamma\left(X;\ \mathscr{C}_p(\mathscr{H}_0(\mathscr{L}^q))\right) = \prod_{i_0,\,\ldots,\,i_p} \Gamma_{M_{i_0}\ldots i_p}\left(X;\ \mathscr{H}_0(\mathscr{L}^q)\right),$$

where

$$0 \to \mathcal{F} \to \mathcal{L}^0 \to \mathcal{L}^1 \to \cdots$$

is a locally fine resolution of the analytic sheaf $\mathcal{F}$ on X. For the first spectral sequence of the double complex $(K_{p,q})$ we get

$$'E^1_{p,q} = \prod_{i_0, \ldots, i_p} H_q^{M_{i_0} \cdots i_p}(X; \mathcal{F}).$$

Consequently,

$$'E^2_{p,q} = H_p(\mathfrak{M}; \mathcal{H}_q^X(\mathcal{F})),$$

where $\mathcal{H}_q^X(\mathcal{F})$ is the copresheaf $M \mapsto H_q M(X; \mathcal{F})$ on the category of closed sets in X. For the second spectral sequence

$$''E^1_{p,q} = 0 \quad \text{for} \quad q \neq 0,$$
$$''E^1_{p,0} = \Gamma(X; \mathcal{H}_0(\mathcal{L}^p)),$$

since $\mathcal{H}_0(\mathcal{L}^p)$ are flabby sheaves (by Lemma 2.1 of Chapter 2). Consequently, the second spectral sequence degenerates:

$$''E^2_{p,q} = 0 \quad \text{for} \quad q \neq 0;$$
$$''E^2_{p,0} = H_p(X; \mathcal{F}).$$

The theorem is proved.

2. Cohomology with Arbitrary Supports

2.1. Let X be a complex space, countable at infinity, $\mathcal{F}$ be a coherent analytic sheaf on X, $\mathfrak{U} = (U_i)$ be a locally finite covering of the space X by holomorphically complete open sets.

2.1.1. Lemma. There exist natural isomorphisms of vector spaces

$$H^k(X; \mathcal{F}) = H^k(\mathfrak{U}; \mathcal{F}).$$

Proof. We make use of the Leray spectral sequence

$$E_2^{p,q} = H^p(\mathfrak{U}; \mathcal{H}^q(\mathcal{F})) \Rightarrow H^{p+q}(X; \mathcal{F})$$

(cf. point 1.9.3 of Chapter 1). By Cartan's theorem (B)

$$H^q(U_{i_0 \ldots i_p}; \mathscr{F}) = 0 \quad \text{for} \quad q \neq 0$$

(cf. point 1.13 of Chapter 1). Consequently, the spectral sequence degenerates and the natural linear map of vector spaces

$$H^k(\mathfrak{U}; \mathscr{F}) \to H^k(X; \mathscr{F})$$

is an isomorphism. The lemma is proved.

2.1.2. We endow each vector space of sections $\Gamma(U_{i_0 \ldots i_k}; \mathscr{F})$ with its natural locally convex topology (cf. Gunning and Rossi [1, p. 303]; cf. also point 3.1 of Chapter 1),[2] which is the topology of a Frechet–Schwartz space. We endow the vector space of cochains $C^k(\mathfrak{U}; \mathscr{F})$ (cf. point 1.9 of Chapter 1) with the product topology, which is also the topology of a Frechet–Schwartz space. We endow the cohomology vector space $H^k(\mathfrak{U}; \mathscr{F})$ with the topology of a quotient-space. Finally, by means of the isomorphism of Lemma 2.1.1 we define a topology on the cohomology vector space $H^k(X; \mathscr{F})$. In view of Lemma 3.6 of Chapter 1, this topology is actually independent of the choice of covering $\mathfrak{U}$. The associated separated topological vector space $\tilde{H}^k(X; \mathscr{F})$ is a Frechet–Schwartz space (the tilde denotes factorization by the closure of zero).

2.2. Let X be a complex space, countable at infinity, $\mathscr{F}$ be a coherent analytic sheaf on X, $\mathfrak{U} = (U_i)$ be a sufficiently fine locally finite covering of the space X by holomorphically complete open sets.

2.2.1. Lemma. There exist natural isomorphisms of vector spaces[3]

$$H_k^c(X; \mathscr{F}) = H_k^c(\mathfrak{U}; \mathscr{H}_0^c(\mathscr{F})).$$

Proof. We make use of the spectral sequence of Theorem 1.1:

$$E_{p,q}^2 = H_p^c(\mathfrak{U}; \mathscr{H}_q^c(\mathscr{F})) \Rightarrow H_{p+q}^c(X; \mathscr{F}).$$

We show that

$$H_q^c(U_{i_0 \ldots i_p}; \mathscr{F}) = 0 \quad \text{for} \quad q \neq 0.$$

One can assume that $U_{i_0 \ldots i_p} = M$ is an analytic set in the domain of holomorphy G of the space $\mathbf{C}^n$. Then there is an isomorphism

$$H_q^c(M; \mathscr{F}) = \operatorname{Ext}_{\mathfrak{O}_G, c}^{n-q}(G; \mathscr{F}^G, \Omega_G^n)$$

(cf. Corollary 3.4 of Chapter 2), and, on the other hand, by Cartan's theorem (B)

$$\text{Ext}^k_{\mathcal{O}_G,\, c}(G;\ \mathscr{F}^G,\ \Omega^n_G) = 0 \quad \text{for}\quad k \neq n$$

(cf. Corollary 3.5.2 of Chapter 1). Thus, the spectral sequence degenerates and, consequently, the natural linear map of vector spaces

$$H^c_k(X;\ \mathscr{F}) \to H^c_k(\mathfrak{U};\ \mathscr{H}^c_0(\mathscr{F}))$$

is an isomorphism. The lemma is proved.

2.2.2. We endow each vector space of zero-dimensional homology $H_0^c(U_{i_0\ldots i_k};\ \mathscr{F})$ with its natural locally convex topology. For this we shall assume that $U_{i_0\ldots i_k} = M$ is an analytic set in the domain G of the space $\mathbf{C}^n$. We make use of the natural isomorphism of vector spaces

$$H^c_0(M;\ \mathscr{F}) = \text{Ext}^n_{\mathcal{O}_G,\, c}(G;\ \mathscr{F}^G,\ \Omega^n_G)$$

(cf. Corollary 3.4 of Chapter 2) and the natural isomorphism of topological vector spaces

$$\text{Ext}^n_{\mathcal{O}_G,\, c}(G;\ \mathscr{F}^G,\ \Omega^n_G) = \{\Gamma(M;\ \mathscr{F})\}'$$

(cf. Corollary 3.5.2 of Chapter 1). By means of these isomorphisms we define a locally convex topology on the vector space $H_0^c(M;\ \mathscr{F})$, which is the topology of the strong dual to a Frechet–Schwartz space. The vector space of chains $C_k^c(\mathfrak{U};\ \mathscr{H}_0^c(\mathscr{F}))$ (cf. point 1.1) will be endowed with the topology of a locally convex direct sum, which is also the topology of a strong dual to a Frechet–Schwartz space. We endow the homology vector space $H_k^c(\mathfrak{U};\ \mathscr{H}_0^c(\mathscr{F}))$ with the quotient topology. And, finally, by means of the isomorphism of Lemma 2.2.1 we define a topology on the homology vector space $H_k^c(X;\ \mathscr{F})$. In fact, in view of Lemma 3.6 of Chapter 1 this topology is independent of the choice of covering $\mathfrak{U}$. The associated separated topological vector space $\widetilde{H}_k^c(X;\ \mathscr{F})$ is the strong dual to a Frechet–Schwartz space (the tilde denotes factorization by the closure of zero).

2.3. Theorem. Let X be a complex space, countable at infinity, and $\mathscr{F}$ be a coherent analytic sheaf on X. Then the topological vector space

$H_k{}^c(X; \mathscr{F})$ is naturally isomorphic with the strong dual to the topological vector space $\tilde{H}^k(X; \mathscr{F})$[4]:

$$\{\tilde{H}^k (X; \mathscr{F})\}' = \tilde{H}_k^c (X; \mathscr{F}).$$

Proof. Let $\mathfrak{U} = (U_i)$ be a sufficiently fine locally finite covering of the space X by holomorphically complete open sets. Then the strong dual to the topological vector space $\Gamma(U_{i_0 \ldots i_k}; \mathscr{F})$ is naturally isomorphic to the topological vector space $H_0{}^c(U_{i_0 \ldots i_k}; \mathscr{F})$ (cf. Corollary 3.5.2 of Chapter 1). Thus, the strong dual to the topological vector space of cochains $C^k(\mathfrak{U}; \mathscr{F})$ (cf. point 1.9 of Chapter 1) is naturally isomorphic to the topological vector space of chains $C_k{}^c(\mathfrak{U}; \mathscr{H}_0{}^c(\mathscr{F}))$ (cf. point 1.1). The linear map, dual to the coboundary operator

$$\delta \colon C^k(\mathfrak{U}; \mathscr{F}) \to C^{k+1}(\mathfrak{U}; \mathscr{F})$$

(cf. point 1.9 of Chapter 1), can be naturally identified with the boundary operator

$$\partial \colon C_{k+1}^c (\mathfrak{U}; \mathscr{H}_0^c (\mathscr{F})) \to C_k^c (\mathfrak{U}; \mathscr{H}_0^c (\mathscr{F}))$$

(cf. point 5.1 of Chapter 1). With the help of Lemma 3.3 of Chapter 1 we get a natural isomorphism of topological vector spaces

$$\{\tilde{H}^k (\mathfrak{U}; \mathscr{F})\}' = \tilde{H}_k^c (\mathfrak{U}; \mathscr{H}_0^c (\mathscr{F}))$$

(the tilde denotes factorization by the closure of zero). The assertion follows from this by virtue of what was recounted in points 2.1.2 and 2.2.2. The theorem is proved.[5]

Corollary. The cohomology topological vector space $H^k(X; \mathscr{F})$ is separable if and only if the homology topological vector space $H_{k-1}{}^c(X; \mathscr{F})$ is separable.

3. Cohomology with Compact Supports

3.1. Let X be a complex space, countable at infinity, $\mathscr{F}$ be a coherent analytic sheaf on X, $\mathfrak{M} = (M_i)$ be a locally finite covering of the space X by compact sets, for each of which there exists a fundamental system of holomorphically complete open neighborhoods. We consider the groups of cochains with compact supports of the covering $\mathfrak{M}$ with coefficients in the sheaf $\mathscr{F}$

$$C_c^k(\mathfrak{M}; \mathscr{F}) = \prod_{i_0, \ldots, i_k} \Gamma(M_{i_0 \ldots i_k}; \mathscr{F}),$$

where the direct sum is taken over those collections of indices $(i_0, \ldots, i_k)$, for which the intersection $M_{i_0 \ldots i_k} = M_{i_0} \cap \ldots \cap M_{i_k}$ is not empty, the coboundary operators

$$\delta \colon C_c^k(\mathfrak{M}; \mathscr{F}) \to C_c^{k+1}(\mathfrak{M}; \mathscr{F})$$

(cf. point 1.9 of Chapter 1), and the cohomology groups with compact supports of the covering $\mathfrak{M}$ with coefficients in the sheaf $\mathscr{F}$:

$$H_c^k(\mathfrak{M}; \mathscr{F}) = H^k C_c^{\bullet}(\mathfrak{M}; \mathscr{F}).$$

3.1.1. Lemma. There exist natural isomorphisms of vector spaces

$$H_c^k(X; \mathscr{F}) = H_c^k(\mathfrak{M}; \mathscr{F}).$$

Proof. We make use of the Leray spectral sequence for a closed covering:

$$E_2^{p, q} = H_c^p(\mathfrak{M}; \ \mathscr{H}^q(\mathscr{F})) \Rightarrow H_c^{p+q}(X; \mathscr{F})$$

(cf. Godement [1, p. 236]), where $\mathscr{H}^q(\mathscr{F})$ is the presheaf $M \mapsto H^q(M; \mathscr{F})$ on the category of closed sets in X. From Cartan's theorem (B) we get

$$H^q(M_{i_0 \ldots i_p}; \mathscr{F}) = 0 \quad \text{for} \quad q \neq 0,$$

so the spectral sequence degenerates. Consequently, the natural linear map of vector spaces

$$H_c^k(\mathfrak{M}; \mathscr{F}) \to H_c^k(X; \mathscr{F})$$

is an isomorphism. The lemma is proved.

3.1.2. For an arbitrary compact set $M \subset X$ we endow the vector space of sections $\Gamma(M; \mathscr{F})$ with the topology of the locally convex inductive limit

$$\Gamma(M; \mathscr{F}) = \lim_{\longrightarrow} \Gamma(U; \mathscr{F})$$

with respect to the filtered set, ordered by inclusion, of all open neighborhoods U of the set M. Then the topological vector space $\Gamma(M; \mathscr{F})$ is the strong dual to a Frechet–Schwartz space (cf. point 3.11 of Chapter 1). We endow the vector space of cochains $C_c^k(\mathfrak{M}; \mathscr{F})$ with the topology of locally convex direct sum, which is also the topology of a strong dual to a Frechet–Schwartz space. The cohomology vector space $H_c^k(\mathfrak{M}; \mathscr{F})$ is endowed with the quotient topology. Thus by means of the isomorphism of Lemma 3.1.1, there is defined a locally convex topology on the cohomology vector space $H_c^k(X; \mathscr{F})$; in view of Lemma 3.6 of Chapter 1, this topology is actually independent of the choice of the covering $\mathfrak{M}$. The associated separated topological vector space $\tilde{H}_c^k(X; \mathscr{F})$ (the tilde denotes factorization by the closure of zero) is the strong dual to a Frechet–Schwartz space.

3.2. Let X be a complex space, countable at infinity, $\mathscr{F}$ be a coherent analytic sheaf on X, $\mathfrak{M} = (M_i)$ be a sufficiently fine locally finite covering of the space X by compact sets, for each of which there exists a fundamental system of holomorphically complete open neighborhoods.

3.2.1. Lemma. There exist natural isomorphisms of vector spaces

$$H_k(X; \mathscr{F}) = H_k(\mathfrak{M}; \mathscr{H}_0^X(\mathscr{F})).$$

Proof. We make use of the spectral sequence of Theorem 1.2:

$$E_{p,q}^2 = H_p(\mathfrak{M}; \mathscr{H}_q^X(\mathscr{F})) \Rightarrow H_{p+q}(X; \mathscr{F}).$$

We show that

$$H_q^{M_{i_0 \cdots i_p}}(X; \mathscr{F}) = 0 \quad \text{for} \quad q \neq 0.$$

One can assume that $M_{i_0 \ldots i_p} = K$ is a compact set in a domain of holomorphy G of the space $\mathbf{C}^n$, having a fundamental system of holomorphically complete open neighborhoods. Then there is a natural isomorphism of vector spaces

$$H_q^K(X; \mathscr{F}) = \operatorname{Ext}_{\mathcal{O}_G, K}^{n-q}(G; \mathscr{F}^G, \Omega_G^n)$$

(cf. Corollary 3.4 of Chapter 2). On the other hand, by Cartan's theorem (B),

$$H^q(K; \mathscr{F}) = 0 \quad \text{for} \quad q \neq 0$$

and, consequently,

$$\operatorname{Ext}_{\mathcal{O}_G, K}^{n-q}(G; \mathscr{F}^G, \Omega_G^n) = 0 \quad \text{for} \quad q \neq 0$$

(cf. Corollary 3.12.2 of Chapter 1). Thus the spectral sequence degenerates; we get that the natural linear map of vector spaces

$$H_k(X; \mathscr{F}) \to H_k(\mathfrak{M}; \mathscr{H}_0^X(\mathscr{F}))$$

is an isomorphism. The lemma is proved.

3.2.2. We endow each vector space of zero-dimensional homology $H_0^{M_{i_0 \ldots i_k}}(X; \mathscr{F})$ with its natural locally convex topology. For this we shall assume that $M_{i_0 \ldots i_k} = K$ is a compact set in a domain of holomorphy G of the space $\mathbf{C}^n$, having a fundamental system of holomorphically complete open neighborhoods. We make use of the natural isomorphism of vector spaces

$$H_0^K(X; \mathscr{F}) = \mathrm{Ext}_{\mathscr{O}_G, K}^n(G; \mathscr{F}^G, \Omega_G^n)$$

(cf. Corollary 3.4 of Chapter 2) and the natural isomorphism of topological vector spaces

$$\mathrm{Ext}_{\mathscr{O}_G, K}^n(G; \mathscr{F}^G, \Omega_G^n) = \{\Gamma(K; \mathscr{F})\}'$$

(cf. Corollary 3.12.2 of Chapter 1). With the help of these isomorphisms we define a locally convex topology in the vector space $H_0^K(X; \mathscr{F})$, which is the topology of a Frechet–Schwartz space. We endow the vector space of chains $C_k(\mathfrak{M}; \mathscr{H}_0^X(\mathscr{F}))$ (cf. point 1.2) with the product topology, which is the topology of a Frechet–Schwartz space. We endow the homology vector space $H_k(\mathfrak{M}; \mathscr{H}_0^X(\mathscr{F}))$ with the quotient topology. By means of the isomorphism of Lemma 3.2.1 we thus define a locally convex topology on the cohomology vector space $H_k(X; \mathscr{F})$; in view of Lemma 3.6 of Chapter 1, this topology is actually independent of the choice of the covering $\mathfrak{M}$. The associated separated topological vector space $\tilde{H}_k(X; \mathscr{F})$ (the tilde denotes factorization by the closure of zero) is a Frechet–Schwartz space.

3.3. Theorem. Let X be a complex space, countable at infinity, and $\mathscr{F}$ be a coherent analytic sheaf on X. Then the topological vector space $\tilde{H}_k(X; \mathscr{F})$ is naturally isomorphic with the strong dual to the topological vector space $\tilde{H}_c^k(X; \mathscr{F})$[6]:

$$\{\tilde{H}_k(X; \mathscr{F})\}' = \tilde{H}_c^k(X; \mathscr{F}).$$

Proof. Let $\mathfrak{M} = (M_i)$ be a sufficiently fine locally finite covering of the space X by compact sets, for each of which there exists a fundamental system of holomorphically complete open neighborhoods. Then the strong dual to the topological vector space $\Gamma(M_{i_0\ldots i_k}; \mathscr{F})$ is naturally isomorphic to the topological vector space $H_0^{M io\ldots i_k}(X; \mathscr{F})$ (cf. Corollary 3.12.3 of Chapter 1). The strong dual to the topological vector space of cochains $C_c^k(\mathfrak{M}; \mathscr{F})$ (cf. point 3.1) is naturally isomorphic to the topological vector space of chains $C_k(\mathfrak{M}; \mathscr{H}_0^X(\mathscr{F}))$ (cf. point 1.2). The linear map dual to the coboundary operator

$$\delta: C_c^k(\mathfrak{M}; \mathscr{F}) \rightarrow C_c^{k+1}(\mathfrak{M}; \mathscr{F})$$

(cf. point 1.9 of Chapter 1) can be naturally identified with the boundary operator

$$\partial: C_{k+1}(\mathfrak{M}; \mathscr{H}_0^X(\mathscr{F})) \rightarrow C_k(\mathfrak{M}; \mathscr{H}_0^X(\mathscr{F}))$$

(cf. point 5.1 of Chapter 1). Thus we get a natural isomorphism of topological vector spaces

$$\{\tilde{H}_c^k(\mathfrak{M}; \mathscr{F})\}' = \tilde{H}_k(\mathfrak{M}; \mathscr{H}_0^X(\mathscr{F}))$$

(cf. Lemma 3.3 of Chapter 1). By virtue of what was said in points 3.1.2 and 3.2.2, the assertion of Theorem 3.3 follows directly from this. The theorem is proved.[7]

Corollary. The cohomology topological vector space $H_c^k(X; \mathscr{F})$ is separated if and only if the homology topological vector space $H_{k-1}(X; \mathscr{F})$ is separated.

4. Holomorphically Complete Spaces

4.1. Theorem. If the complex space X is holomorphically complete, then for any coherent analytic sheaf $\mathscr{F}$ on X

$$H_k^c(X; \mathscr{F}) = 0 \quad \text{for} \quad k \neq 0,$$

and the topological vector space $H_0^c(X; \mathscr{F})$ is separated and naturally isomorphic to the strong dual to the topological vector space $\Gamma(X; \mathscr{F})$[8]:

$$H_0^c(X; \mathscr{F}) = \{\Gamma(X; \mathscr{F})\}'.$$

Proof. By Cartan's theorem (B),

$$H^k(X; \mathcal{F}) = 0 \quad \text{for} \quad k \neq 0.$$

Consequently, the topological vector spaces $H_k{}^c(X; \mathcal{F})$ are separated for all k (cf. the corollary to Theorem 2.3). To complete the proof it suffices to apply Theorem 2.3. Theorem 4.1 is proved.

4.2. Lemma. Let X be a complex space, countable at infinity, and U be an arbitrary open set in X. Then for any coherent analytic sheaf $\mathcal{F}$ on X the natural map

$$H_c^k(U; \mathcal{F}) \to H_c^k(X; \mathcal{F}),$$

defined by the inclusion $U \subset X$, is continuous.

Proof. Let $\mathfrak{M} = (M_i)$ be a locally finite covering of the open set U by compact sets, for each of which there exists a fundamental system of holomorphically complete open neighborhoods. For each compact set $S \subset U$ we set

$$C_S^k(\mathfrak{M}; \mathcal{F}) = \prod_{i_0 \cdots i_k} \Gamma(M_{i_0 \cdots i_k}; \mathcal{F}),$$

where the direct sum is taken over those collections of indices $(i_0, \ldots, i_k)$, for which $M_{i_0 \ldots i_k} \subset S$. Then there is a natural isomorphism of topological vector spaces, strongly dual to Frechet–Schwartz spaces:

$$C_c^k(\mathfrak{M}; \mathcal{F}) = \lim_{\longrightarrow} C_S^k(\mathfrak{M}; \mathcal{F}),$$

where the inductive limit is taken with respect to the filtered set, ordered by inclusion, of all compact subsets $S \subset U$. From this we get a natural isomorphism of topological vector spaces

$$H_c^k(\mathfrak{M}; \mathcal{F}) = \lim_{\longrightarrow} H_S^k(\mathfrak{M}; \mathcal{F}).$$

Obviously for each compact set $S \subset U$ the natural map

$$H_S^k(\mathfrak{M}; \mathcal{F}) \to H_c^k(X; \mathcal{F})$$

is continuous (cf. Lemma 3.1.1). Hence the natural map $H_c{}^k(U; \mathcal{F}) \to H_c{}^k(X; \mathcal{F})$ is also continuous. The lemma is proved.

Corollary. The restriction map

$$H_k(X; \mathcal{F}) \to H_k(U; \mathcal{F})$$

is continuous.

Proof. Passing to dual spaces in Lemma 4.2, by Theorem 3.3 we get that the restriction map

$$\tilde{H}_k(X; \mathscr{F}) \to \tilde{H}_k(U; \mathscr{F})$$

is continuous. The assertion follows directly from this.

4.3. Lemma. Let M be an analytic set in the domain G of the space $\mathbf{C}^n$ and $\mathscr{F}$ be a coherent analytic sheaf on M, for which on G there is a free resolution

$$0 \to 6_G^{s_n} \xrightarrow{\sigma_n} \ldots \xrightarrow{\sigma_2} 6_G^{s_1} \xrightarrow{\sigma_1} 6_G^{s_0} \to \mathscr{F}^G \to 0.$$

Then there exists a natural isomorphism of topological vector spaces

$$H_c^k(M; \mathscr{F}) = \operatorname{Ext}_{6_G, c}^k(G; \Omega_G^n, \mathscr{F}^G \otimes_{6_G} \Omega_G^n).$$

Proof. Let $\mathfrak{M} = (M_i)$ be a locally finite covering of the domain G by compact sets, for each of which there exists a fundamental system of holomorphically complete open neighborhoods. We consider the double complex

$$K_c^{p, q}(\mathfrak{M}) = \prod_{j-i=q} C_c^p(\mathfrak{M}; \mathscr{H}om_{6_G}(\Omega_G^n, 6_G^{s_i} \otimes_{6_G} {}'\mathscr{D}_G^{n, l})),$$

whose differentials are defined by the homomorphisms σ_i, which are the differentials of the Dolbeault–Grothendieck resolution

$$0 \to \Omega_G^n \to {}'\mathscr{D}_G^{n, 0} \xrightarrow{d''} {}'\mathscr{D}_G^{n, 1} \xrightarrow{d''} \ldots \xrightarrow{d''} {}'\mathscr{D}_G^{n, n} \to 0$$

and the coboundary operator δ corresponding to the covering $\mathfrak{M}$. We denote by $K_c^{\cdot}(\mathfrak{M})$ the associated single complex:

$$K_c^r(\mathfrak{M}) = \prod_{p+q=r} K_c^{p, q}(\mathfrak{M}).$$

For the first filtration of the complex $K_c^{\cdot}(\mathfrak{M})$ we get

$$'E_1^{p, q} = 0 \quad \text{for} \quad q \neq 0;$$
$$'E_1^{p, 0} = C_c^p(\mathfrak{M}; \mathscr{F}^G).$$

Consequently, the first spectral sequence degenerates:

$$'E_2^{p,\,q} = 0 \quad \text{for} \quad q \neq 0;$$
$$'E_2^{p,\,0} = H_c^p(M;\ \mathcal{F}).$$

We consider the complex

$$K^r = \prod_{j-i=r} C_c^j\left(\mathfrak{M};\ \mathcal{H}om_{\mathcal{O}_G}(\Omega_G^n,\ \mathcal{O}_G^{s_j} \otimes_{\mathcal{O}_G} \Omega_G^n)\right).$$

Then there exists a natural continuous linear morphism of complexes

$$K^r \to K_c^r(\mathfrak{M}), \tag{1}$$

which is an algebraic quasiisomorphism. On the other hand, there exists a natural continuous linear morphism of complexes

$$K^r \to {'E_1^{r,\,0}}, \tag{2}$$

which is also an algebraic quasiisomorphism. For the second filtration of the complex $K_c^{\cdot}(\mathfrak{M})$ we get

$$''E_1^{p,\,q} = 0 \quad \text{for} \quad q \neq 0;$$
$$''E_1^{p,\,0} = \prod_{j-i=p} \operatorname{Hom}_{\mathcal{O}_{G,\ c}}(G;\ \Omega_G^n,\ \mathcal{O}_G^{s_j} \otimes_{\mathcal{O}_G} {'\mathcal{D}_G^{n,\,i}}).$$

Consequently, the second spectral sequence also degenerates:

$$''E_2^{p,\,q} = 0 \quad \text{for} \quad q \neq 0;$$
$$''E_2^{p,\,0} = \operatorname{Ext}_{\mathcal{O}_{G,\ c}}^p(G;\ \Omega_G^n,\ \mathcal{F}^G \otimes_{\mathcal{O}_G} \Omega_G^n)$$

(cf. Lemma 3.2 of Chapter 1). Thus, the natural continuous linear morphism of complexes

$$''E_1^{r,\,0} \to K_c^r(\mathfrak{M}) \tag{3}$$

is an algebraic quasiisomorphism. By Lemma 3.6 of Chapter 1, the morphisms of complexes (1)–(3) are topological quasiisomorphisms. Thus they define isomorphisms of topological vector spaces

$$'E_2^{r,\,0} = {''E_2^{r,\,0}}.$$

The lemma is proved.

Corollary. There exists a natural isomorphism of topological vector spaces

$$H_k(M; \mathcal{F}) = \text{Ext}_{\mathcal{O}_G}^{n-k}(G; \mathcal{F}^G, \Omega_G^n).$$

Proof. Passing in Lemma 4.3 to dual spaces, by Theorem 3.4 of Chapter 1 and Theorem 3.3 of Chapter 3, we get that the natural algebraic isomorphism between $H_k(M; \mathcal{F})$ and $\text{Ext}_{\mathcal{O}_G}{}^{n-k}(G; \mathcal{F}^G, \Omega_G^n)$, which exists by virtue of the corollary to Proposition 3.4 of Chapter 2, induces an isomorphism of the associated separable topological vector spaces

$$\tilde{H}_k(M; \mathcal{F}) = \widetilde{\text{Ext}}_{\mathcal{O}_G}^{n-k}(G; \mathcal{F}^G, \Omega_G^n).$$

The assertion follows directly from this.

4.4. Theorem. If the complex space X is holomorphically complete, then for any coherent analytic sheaf $\mathcal{F}$ on X the topological vector space $H_k(X; \mathcal{F})$ is separated and naturally isomorphic to the strong dual to the separated topological vector space $H_c{}^k(X; \mathcal{F})$[9]:

$$\{H_c^k(X; \mathcal{F})\}' = H_k(X; \mathcal{F}).$$

Proof. Let U be an arbitrary holomorphically complete open set in X. We make use of the spectral sequence

$$E_{p,q}^2 = H^{-p}(U; \mathcal{H}_q(\mathcal{F})) \Rightarrow H_{p+q}(U; \mathcal{F})$$

(cf. Theorem 4.1 of Chapter 2). Since the sheaves $\mathcal{H}_q(\mathcal{F})$ are coherent, by Cartan's theorem (B) the spectral sequence degenerates:

$$\begin{aligned} E_{p,q}^2 &= 0 \quad \text{for} \quad p \neq 0, \\ E_{0,q}^2 &= \Gamma(U; \mathcal{H}_q(\mathcal{F})). \end{aligned}$$

Consequently, the natural linear map

$$H_k(U; \mathcal{F}) \longrightarrow \Gamma(U; \mathcal{H}_k(\mathcal{F}))$$

is an isomorphism of vector spaces. Let $\mathcal{U} = (U_i)$ be a sufficiently fine locally finite covering of the space X by holomorphically complete open sets. Then the natural map

$$H_k(X; \mathcal{F}) \longrightarrow \prod_i H_k(U_i; \mathcal{F}),$$

defined by the restrictions, is injective and continuous (cf. the corollary to Lemma 4.2). On the other hand, by the corollary to Lemma 4.3, the spaces $H_k(U_i; \mathcal{F})$ are separated (cf. Theorem 3.7 of Chapter 1), so the spaces $H_k(X; \mathcal{F})$ are separated, and then by the corollary to Theorem 3.3 the spaces $H_c^k(X; \mathcal{F})$ are also separated. The theorem is proved.

4.5. Lemma. Let $u: E \to F$ be a continuous linear map of Frechet spaces. Then the quotient space $F/u(E)$ is separated, if it is finite-dimensional.[10]

Proof. Let M be an algebraic complement of the subspace $u(E)$ in F. Then the map

$$v: E \times M \to F,$$

for which $v(x, y) = u(x) + y$ for $x \in E$ and $y \in M$, is surjective, and by Banach's theorem is a homomorphism. Consequently, u is also a homomorphism, and its image $u(E)$ is closed in F. The lemma is proved.

4.6. Theorem. If the complex space X is compact, then for any coherent analytic sheaf $\mathcal{F}$ on X, the topological vector spaces $H^k(X; \mathcal{F})$ and $H_k(X; \mathcal{F})$ are separated, finite-dimensional, and naturally dual to one another.[11]

Proof. By the Cartan–Serre finiteness theorem (cf. point 1.15.2 of Chapter 1) the topological vector spaces $H^k(X; \mathcal{F})$ are finite-dimensional. By Lemma 4.5 they are separated, and hence the spaces $H_k(X; \mathcal{F})$ are also separated (cf. the corollary to Theorem 2.3). Thus the assertion follows from Theorem 2.3. The theorem is proved.[12]

5. Leray Spectral Sequence

5.1. Lemma. Let $f: X \to Y$ be a proper morphism of complex spaces (cf. point 1.15 of Chapter 1) and U be a relatively compact holomorphically complete open set in Y. Then for any coherent analytic sheaf $\mathcal{F}$ on X there is a natural isomorphism of topological vector spaces

$$H^q(f^{-1}(U); \mathcal{F}) = \Gamma(U; R^q f_* \mathcal{F}).$$

Proof. We make use of the Leray spectral sequence for cohomology

$$E_2^{p,\,q} = H^p\,(U;\ R^q f_* \mathcal{F}) \Rightarrow H^{p+q}\,(f^{-1}(U);\ \mathcal{F})$$

(cf. point 1.14.2 of Chapter 1). By Grauert's theorem on direct images the sheaves $R^q f_* \mathcal{F}$ are coherent (cf. point 1.15.1 of Chapter 1). Consequently, the Leray spectral sequence degenerates. Hence the natural linear map

$$H^q\,(f^{-1}(U);\ \mathcal{F}) \to \Gamma\,(U;\ R^q f_* \mathcal{F}) \tag{1}$$

is an isomorphism of vector spaces. We endow the vector spaces $H^q(f^{-1}(U);\ \mathcal{F})$ and $\Gamma(U;\ R^q f_* \mathcal{F})$ with their usual topologies. Since the sheaf $R^q f_* \mathcal{F}$ is generated over U by a finite number of its sections, the topology of the space $\Gamma(U;\ R^q f_* \mathcal{F})$ is the strongest of the locally convex topologies compatible with the structure of $\Gamma(U;\ \mathcal{O}_Y)$-module. It follows directly from this that the map inverse to the isomorphism (1) is continuous. We show that the topological vector space $H^q(f^{-1}(U);\ \mathcal{F})$ is separated. Then by Banach's closed graph theorem the linear map (1) will be an isomorphism of topological vector spaces and the proof of Lemma 5.1 will be finished.

For an arbitrary point $y \in U$ let $\mathcal{J}_y$ be the coherent subsheaf of ideals in $\mathcal{O}_X$, generated by functions of the form $\varphi \circ f$, where $\varphi \in \Gamma(V;\ \mathcal{O}_Y)$ for some open set $V \subset Y$ and $\varphi(y) = 0$, if $y \in V$. Then for each $k = 1, 2, \ldots$ we get a coherent analytic sheaf on X:

$$\mathcal{F}/\mathcal{J}_y^k \mathcal{F} = \mathcal{F} \otimes_{\mathcal{O}_X} (\mathcal{O}_X/\mathcal{J}_y^k).$$

Since in the isomorphism (1) the coherent analytic sheaf $\mathcal{F}$ is arbitrary, we get natural isomorphisms of vector spaces

$$H^q\,(f^{-1}(U);\ \mathcal{F}/\mathcal{J}_y^k \mathcal{F}) = \Gamma\,(U;\ R^q f_* (\mathcal{F}/\mathcal{J}_y^k \mathcal{F})). \tag{2}$$

The sheaf $\mathcal{F}/\mathcal{J}_y^k \mathcal{F}$ is trivial outside the set $f^{-1}(y)$, so the sheaf $R^q f_* (\mathcal{F}/\mathcal{J}_y^k \mathcal{F})$ is trivial everywhere outside the point y. Consequently, if V is a relatively compact, holomorphically complete open set in U such that $y \in V$ and $\bar{V} \subset U$, then the restriction map

$$\Gamma\,(U;\ R^q f_* (\mathcal{F}/\mathcal{J}_y^k \mathcal{F})) \to \Gamma\,(V;\ R^q f_* (\mathcal{F}/\mathcal{J}_y^k \mathcal{F}))$$

is completely continuous and is an isomorphism of topological vector spaces. By Riesz's theorem the vector space $\Gamma(U;\ R^q f_* (\mathcal{F}/\mathcal{J}_y^k \mathcal{F}))$ is finite-dimensional. From this, in view of the isomorphism (2), we get that the topological vector space $H^q(f^{-1}(U); \mathscr{F}/\mathscr{I}_y^k\mathscr{F})$ is separated and finite-dimensional (cf. Lemma 4.5). By Grauert's theorem the natural map

$$\Gamma(U; R^q f_* \mathscr{F}) \to \prod_{\substack{y \in U \\ k \geqslant 1}} \Gamma(U; R^q f_*(\mathscr{F}/\mathscr{I}_y^k\mathscr{F}))$$

(cf. Grauert [4, p. 292]) is injective. Then in view of the isomorphisms (1) and (2) the natural continuous linear map

$$H^q(f^{-1}(U); \mathscr{F}) \to \prod_{\substack{y \in U \\ k \geqslant 1}} H^q(f^{-1}(U); \mathscr{F}/\mathscr{I}_y^k\mathscr{F})$$

is also injective. Thus we get that the topological vector space $H^q(f^{-1}(U); \mathscr{F})$ is separated. The lemma is proved.

5.2. Corollary. Let $f: X \to Y$ be a proper morphism of complex spaces and U be a relatively compact holomorphically complete open set in Y. Then for any coherent analytic sheaf $\mathscr{F}$ on X there is a natural isomorphism of topological vector spaces

$$H_q^c(f^{-1}(U); \mathscr{F}) = H_0^c(U; R^q f_* \mathscr{F}).$$

Proof. Passing in Lemma 5.1 to the dual spaces, by Theorem 2.3 we get a natural isomorphism of topological vector spaces

$$\tilde{H}_q^c(f^{-1}(U); \mathscr{F}) = \tilde{H}_0^c(U; R^q f_* \mathscr{F}).$$

On the other hand, by the corollary to Theorem 2.3, the topological vector spaces $H_q^c(f^{-1}(U); \mathscr{F})$ and $H_0^c(U; R^q f_* \mathscr{F})$ are separated. Thus the assertion is proved.

5.3. Lemma. Let X be a complex space, which is countable at infinity, and $\mathfrak{U} = (U_i)$ be a locally finite covering of the space X by relatively compact open sets. Then for any analytic sheaf $\mathscr{F}$ on X there exists a spectral sequence

$$E_{p,q}^2 = H_p(\mathfrak{U}; \mathscr{H}_q^c(\mathscr{F})) \Rightarrow H_{p+q}(X; \mathscr{F}),$$

where $\mathscr{H}_q^c(\mathscr{F})$ is the copresheaf $U \to H_q^c(U; \mathscr{F})$ on the category of open sets in X.[13]

Proof. For an arbitrary analytic sheaf $\mathscr{L}$ on X we set

$$\mathscr{C}_p(\mathscr{L}) = \prod_{i_0, \ldots, i_p} \mathscr{L}_{U_{i_0 \ldots i_p}}.$$

We get an exact sequence of analytic sheaves

$$\ldots \xrightarrow{\partial} \mathscr{C}_1(\mathscr{L}) \xrightarrow{\partial} \mathscr{C}_0(\mathscr{L}) \to \mathscr{L} \to 0$$

(cf. the proof of Theorem 1.1). We consider the double complex

$$K_{p, q} = \Gamma(X; \mathscr{C}_p(\mathscr{H}_0(\mathscr{L}^q))) = \prod_{i_0, \ldots, i_p} \Gamma_c(U_{i_0 \ldots i_p}; \mathscr{H}_0(\mathscr{L}^q)),$$

where $\mathscr{L}^{\cdot}$ is a locally fine resolution of the analytic sheaf $\mathscr{F}$. For the first spectral sequence we get

$$'E^1_{p, q} = \prod_{i_0, \ldots, i_p} H^c_q(U_{i_0 \ldots i_p}; \mathscr{F}),$$

whence

$$'E^2_{p, q} = H_p(\mathfrak{U}; \mathscr{H}^c_q(\mathscr{F})).$$

For the second spectral sequence

$$''E^1_{p, q} = 0 \quad \text{for} \quad q \neq 0,$$
$$''E^1_{p, 0} = \Gamma(X; \mathscr{H}_0(\mathscr{L}^p)),$$

since the sheaves $\mathscr{H}_0(\mathscr{L}^p)$ are flabby. Thus, the second spectral sequence degenerates:

$$''E^2_{p, q} = 0 \quad \text{for} \quad q \neq 0;$$
$$''E^2_{p, 0} = H_p(X; \mathscr{F}).$$

The lemma is proved.

5.4. Corollary. Let X be a complex space which is countable at infinity, $\mathscr{F}$ be a coherent analytic sheaf on X, $\mathfrak{U} = (U_i)$ be a locally finite covering of the space X by holomorphically complete open sets. Then there is a natural isomorphism of vector spaces

$$H_k(X; \mathscr{F}) = H_k(\mathfrak{U}; \mathscr{H}^c_0(\mathscr{F})).$$

Proof. By Theorem 4.1

$$H_q^c(U_{i_0 \ldots i_p}; \; \mathscr{F}) = 0 \quad \text{for} \quad q \neq 0.$$

Consequently, the spectral sequence of Lemma 5.3 degenerates. The assertion is proved.

5.5. Theorem. Let $f: X \to Y$ be a proper morphism of complex spaces which are countable at infinity. Then for any coherent analytic sheaf $\mathscr{F}$ on X there exists a spectral sequence[14]

$$E_{p,q}^2 = H_p(Y; \; R^q f_* \mathscr{F}) \Rightarrow H_{p+q}(X; \; \mathscr{F}) \; .$$

Proof. Let $\mathfrak{U} = (U_i)$ be a locally finite covering of the space Y by relatively compact holomorphically complete open sets. Then $f^{-1}(\mathfrak{U}) = (f^{-1}(U_i))$ is a locally finite covering of the space X by open sets. We choose a locally finite covering $\mathfrak{V} = (V_j)$ of the space X by relatively compact holomorphically complete open sets, inscribed in $f^{-1}(\mathfrak{U})$. We consider the double complex

$$K_{p,q} = \prod_{\substack{i_0, \ldots, i_p \\ j_0, \ldots, j_q}} H_0^c(f^{-1}(U_{i_0 \ldots i_p}) \cap V_{j_0 \ldots j_q}; \; \mathscr{F}).$$

For the first spectral sequence of the double complex $(K_{p,q})$ we get

$${}'E_{p,q}^1 = \prod_{i_0, \ldots, i_p} H_q^c(f^{-1}(U_{i_0 \ldots i_p}); \; \mathscr{F})$$

(cf. Lemma 2.2.1). From this, by virtue of Corollary 5.2 we get a natural isomorphism

$${}'E_{p,q}^1 = \prod_{i_0, \ldots, i_p} H_0^c(U_{i_0 \ldots i_p}; \; R^q f_* \mathscr{F}).$$

Thus,

$${}'E_{p,q}^2 = H_p(Y; \; R^q f_* \mathscr{F})$$

(cf. Corollary 5.4). Since the covering $\mathfrak{V}$ is inscribed in the covering $f^{-1}(\mathfrak{U})$, the sequence

$$0 \to \Gamma(V_{j_0 \ldots j_p}; \; \mathscr{F}) \to C^0(f^{-1}(\mathfrak{U}) \cap V_{j_0 \ldots j_p}; \; \mathscr{F}) \xrightarrow{\delta} \ldots$$

(the Aleksandrov–Čech complex) is exact. Passing to dual spaces, we get an exact sequence

$$\ldots \xrightarrow{\partial} C_0^c(f^{-1}(\mathfrak{U}) \cap V_{i_0 \ldots i_p};\ \mathcal{H}_0^c(\mathcal{F})) \rightarrow H_0^c(V_{i_0 \ldots i_p};\ \mathcal{F}) \rightarrow 0$$

(cf. Corollary 3.5.2 of Chapter 1). Thus, for the second spectral sequence of the double complex $(K_{p,q})$ we get

$$''E_{p,q}^1 = 0 \quad \text{for} \quad q \neq 0;$$
$$''E_{p,0}^1 = \prod_{i_0, \ldots, i_p} H_0^c(V_{i_0 \ldots i_p};\ \mathcal{F}).$$

Consequently, the second spectral sequence degenerates:

$$''E_{p,q}^2 = 0 \quad \text{for} \quad q \neq 0;$$
$$''E_{p,0}^2 = H_p(X;\ \mathcal{F})$$

(cf. Corollary 5.4). The theorem is proved.

5.6. Corollary. Let $f: X \rightarrow Y$ be a proper morphism of complex spaces. Then for any coherent analytic sheaf $\mathcal{F}$ on X there exist two convergent spectral sequences, for which the initial terms are respectively equal to

$$'E_{p,q}^2 = \mathcal{H}_p(R^q f_* \mathcal{F}); \quad ''E_{p,q}^2 = R^{-p} f_* \mathcal{H}_q(\mathcal{F}),$$

and the terms $'E^\infty$ and $''E^\infty$ are the bigraded $\mathcal{O}_Y$-modules associated, respectively, with the suitably filtered graded $\mathcal{O}_Y$-module generated by the graded presheaf

$$U \mapsto H(f^{-1}(U);\ \mathcal{F})$$

on the category of open sets of the space Y.[15]

Proof. For an arbitrary open set $U \subset Y$ from Theorem 5.5 we get a spectral sequence

$$'E_{p,q}^2 = H_p(U;\ R^q f_* \mathcal{F}) \Rightarrow H_{p+q}(f^{-1}(U);\ \mathcal{F}).$$

On the other hand, by Theorem 4.1 of Chapter 2 there exists a spectral sequence

$$''E_{p,q}^2 = H^{-p}(f^{-1}(U);\ \mathcal{H}_q(\mathcal{F})) \Rightarrow H_{p+q}(f^{-1}(U);\ \mathcal{F}).$$

The assertion follows from this directly.

Chapter 4

LOCAL HOMOLOGY

1. Homology of a Closed Set

1.1. Let X be a complex space which is countable at infinity, S be a closed set in X, $\mathfrak{U}$ be a locally finite covering of the space X by relatively compact open sets, $\mathfrak{B}$ be a locally finite covering of the set $X \setminus S$ by relatively compact open sets, inscribed in $\mathfrak{U}$. For an arbitrary copresheaf of Abelian groups F on the category of open sets in X we define the cone $Kc_{\bullet}(F)$ of the chain map

$$C^{\bullet}_{\bullet}(\mathfrak{B}; F) \to C^{\bullet}_{\bullet}(\mathfrak{U}; F),$$

by defining its components and differentials in the familiar way (cf. MacLane [1, pp. 67-68]).

Lemma. If $\mathscr{F}$ is a locally fine analytic sheaf on X, then

$$H_k K^c_{\bullet}(\mathscr{H}^c_0(\mathscr{F})) = 0 \quad \text{for} \quad k \neq 0;$$
$$H_0 K^c_{\bullet}(\mathscr{H}^c_0(\mathscr{F})) = \Gamma_c(S; \mathscr{H}_0(\mathscr{F})),$$

where $\mathscr{H}_0{}^c(\mathscr{F})$ is the copresheaf $U \mapsto H_0{}^c(U; \mathscr{F})$ on the category of open sets in X.

Proof. Since $\mathscr{F}$ is a locally fine analytic sheaf on X, by Theorem 2.4 of Chapter 2 the spectral sequence

$$E^2_{p,\,q} = H^c_p(\mathfrak{U}; \mathscr{H}^c_q(\mathscr{F})) \Rightarrow H^c_{p+q}(X; \mathscr{F})$$

109

(cf. Theorem 1.1 of Chapter 3) degenerates. We get a natural isomorphism

$$H_k^c(\mathfrak{U};\ \mathcal{H}_0^c(\mathcal{F})) = H_k^c(X;\ \mathcal{F}).$$

Applying Theorem 2.4 of Chapter 2 again, we get

$$H_k^c(\mathfrak{U};\ \mathcal{H}_0^c(\mathcal{F})) = 0 \quad \text{for} \quad k \neq 0,$$
$$H_0^c(\mathfrak{U};\ \mathcal{H}_0^c(\mathcal{F})) = \Gamma_c(X;\ \mathcal{H}_0(\mathcal{F})).$$

Analogously, we get

$$H_k^c(\mathfrak{B};\ \mathcal{H}_0^c(\mathcal{F})) = 0 \quad \text{for} \quad k \neq 0;$$
$$H_0^c(\mathfrak{B};\ \mathcal{H}_0^c(\mathcal{F})) = \Gamma_c(X \setminus S;\ \mathcal{H}_0(\mathcal{F})).$$

Moreover, by Lemma 2.1 of Chapter 2 there is an exact sequence

$$0 \to \Gamma_c(X \setminus S; \mathcal{H}_0(\mathcal{F})) \to \Gamma_c(X; \mathcal{H}_0(\mathcal{F})) \to \Gamma_c(S; \mathcal{H}_0(\mathcal{F})) \to 0.$$

From this and the exact homology sequence

$$\ldots \to H_k^c(\mathfrak{B};\ F) \to H_k^c(\mathfrak{U};\ F) \to H_k K_\cdot^c(F) \to$$
$$\to H_{k-1}^c(\mathfrak{B};\ F) \to \ldots \to H_0^c(\mathfrak{B};\ F) \to H_0^c(\mathfrak{U};\ F) \to H_0 K_\cdot^c(F) \to 0$$

we get the assertion of Lemma 1.1. The lemma is proved.

1.2. Theorem. Let X be a complex space, countable at infinity, and S be an arbitrary closed set in X. Then for any analytic sheaf $\mathcal{F}$ on X there exists a spectral sequence

$$E_{p,\,q}^2 = H_p K_\cdot^c(\mathcal{H}_q^c(\mathcal{F})) \Rightarrow H_{p+q}^c(S;\ \mathcal{F}),$$

where $\mathcal{H}_q{}^c(\mathcal{F})$ is the copresheaf $U \mapsto H_q{}^c(U;\ \mathcal{F})$ on the category of open sets in X.

Proof. We choose an arbitrary locally fine resolution

$$0 \to \mathcal{F} \to \mathcal{L}^0 \to \mathcal{L}^1 \to \ldots$$

of the analytic sheaf $\mathcal{F}$ on X. We consider the double complex

$$K_{p,\,q} = K_p^c(\mathcal{H}_0^c(\mathcal{L}^q)).$$

For its first spectral sequence we get

$$'E^1_{p,\,q} = K^c_p\,(\mathcal{H}^c_q\,(\mathcal{F})),$$
$$'E^2_{p,\,q} = H_p K^c_.\,(\mathcal{H}^c_q\,(\mathcal{F})).$$

For the second spectral sequence of the double complex $(K_{p,q})$ we get

$$"E^1_{p,\,q} = 0 \quad \text{for} \quad q \neq 0,$$
$$"E^1_{p,\,0} = \Gamma_c\,(S;\ \mathcal{H}_0\,(\mathcal{L}^p))$$

(cf. Lemma 1.1). Consequently, the second spectral sequence degenerates:

$$"E^2_{p,\,q} = 0 \quad \text{for} \quad q \neq 0,$$
$$"E^2_{p,\,0} = H^c_p\,(S;\ \mathcal{F}).$$

The theorem is proved.

2. Local Homology

2.1. Let X be a complex space which is countable at infinity. For each $j = 1, 2, \ldots$ we define by induction a locally finite covering $\mathfrak{M}_j$ of the space X by compact sets such that the following conditions hold:

a) the covering $\mathfrak{M}_j$ consists of sets which are small of order $1/j$ with respect to a fixed metric defining the topology of the space X[1];

b) the covering $\mathfrak{M}_{j+1}$ is the disjoint union of finite coverings of all sets belonging to the covering $\mathfrak{M}_j$.

For each $j = 1, 2, \ldots$ for an arbitrary copresheaf of Abelian groups F on the category of closed sets in X there is defined a natural map of chain complexes

$$C_k\,(\mathfrak{M}_{j+1};\ F) \to C_k\,(\mathfrak{M}_j;\ F)$$

and, consequently, the corresponding projective limit may be defined[2]

$$C^X_k\,(F) = \varprojlim C_k\,(\mathfrak{M}_j;\ F).$$

For an arbitrary closed set $S \subset X$ we set

$$C^{X \setminus S}_k\,(\mathfrak{M}_j;\ F) = \prod_{i_0,\,\ldots,\,i_k} F\,(M_{i_0 \ldots i_k}),$$

where the product is taken over those collections of indices $(i_0, \ldots, i_k)$, for which the intersection $M_{i_0 \ldots i_k}$ of the sets $M_{i_0}, \ldots, M_{i_k} \in \mathfrak{M}_j$ is contained in $X \setminus S$. Analogously to the preceding, there are defined natural maps of chain complexes

$$C_k^{X \setminus S}(\mathfrak{M}_{j+1};\ F) \to C_k^{X \setminus S}(\mathfrak{M}_j;\ F)$$

and, consequently, the corresponding projective limit

$$C_k^{X \setminus S}(F) = \varprojlim C_k^{X \setminus S}(\mathfrak{M}_j;\ F)$$

may be defined.

For each $j = 1, 2, \ldots$ we define the cone $K_{\cdot}(\mathfrak{M}_j;\ F)$ of the chain map

$$C_{\cdot}^{X}(\mathfrak{M}_j;\ F) \to C_{\cdot}^{X \setminus S}(\mathfrak{M}_j;\ F)$$

by giving its components and differentials in the familiar way (cf. MacLane [1, pp. 67-68]). Analogously we define the cone $K_{\cdot}^{S}(F)$ of the chain map

$$C_{\cdot}^{X}(F) \to C_{\cdot}^{X \setminus S}(F).$$

Lemma. If $\mathscr{F}$ is a locally fine analytic sheaf on X, then

$$H_k K_{\cdot}^{S}(\mathscr{H}_0^X(\mathscr{F})) = 0 \quad \text{for} \quad k \neq 0,$$
$$H_0 K_{\cdot}^{S}(\mathscr{H}_0^X(\mathscr{F})) = \Gamma_S(X;\ \mathscr{H}_0(\mathscr{F})),$$

where $\mathscr{H}_0^X(\mathscr{F})$ is the copresheaf $M \mapsto H_0 M(X;\ \mathscr{F})$ on the category of closed sets in X.

Proof. Since $\mathscr{F}$ is a locally fine analytic sheaf on X, by Theorem 2.4 of Chapter 2 the spectral sequence

$$E_{p,\ q}^2 = H_p(\mathfrak{M}_j;\ \mathscr{H}_q^X(\mathscr{F})) \Rightarrow H_{p+q}(X;\ \mathscr{F})$$

(cf. Theorem 1.2 of Chapter 3) degenerates. We get a natural isomorphism

$$H_k(\mathfrak{M}_j;\ \mathscr{H}_0^X(\mathscr{F})) = H_k(X;\ \mathscr{F}).$$

Applying Theorem 2.4 of Chapter 2 again, we get

$$H_k(\mathfrak{M}_j;\ \mathscr{H}_0^X(\mathscr{F})) = 0 \quad \text{for} \quad k \neq 0,$$
$$H_0(\mathfrak{M}_j;\ \mathscr{H}_0^X(\mathscr{F})) = \Gamma(X;\ \mathscr{H}_0(\mathscr{F})).$$

Passing to the projective limit, with the help of the exact sequence

$$0 \to \lim^{(1)} H_{k+1}(\mathfrak{M}_j;\ \mathcal{H}_0^X(\mathcal{F})) \to$$
$$\to H_k C_\bullet^X(\mathcal{H}_0^X(\mathcal{F})) \to \lim H_k(\mathfrak{M}_j;\ \mathcal{H}_0^X(\mathcal{F})) \to 0$$

(cf. Roos [1, p. 3703]), we get

$$H_k C_\bullet^X(\mathcal{H}_0^X(\mathcal{F})) = 0 \quad \text{for} \quad k \neq 0,$$
$$H_0 C_\bullet^X(\mathcal{H}_0^X(\mathcal{F})) = \Gamma(X;\ \mathcal{H}_0(\mathcal{F})).$$

Analogously we get

$$H_k C_\bullet^{X \setminus S}(\mathcal{H}_0^X(\mathcal{F})) = 0 \quad \text{for} \quad k \neq 0,$$
$$H_0 C_\bullet^{X \setminus S}(\mathcal{H}_0^X(\mathcal{F})) = \Gamma(X \setminus S;\ \mathcal{H}_0(\mathcal{F})).$$

Moreover, there is an exact sequence

$$0 \to \Gamma_S(X;\mathcal{H}_0^\cdot(\mathcal{F})) \to \Gamma(X;\mathcal{H}_0(\mathcal{F})) \to \Gamma(X \setminus S;\ \mathcal{H}_0(\mathcal{F})) \to 0$$

(cf. Lemma 2.1 of Chapter 2). From this and from the exact sequence of homology groups

$$\cdots \to H_k K_\bullet^S(F) \to H_k C_\bullet^X(F) \to H_k C_\bullet^{X \setminus S}(F) \to$$
$$\to H_{k-1} K_\bullet^S(F) \to \cdots \to H_0 K_\bullet^S(F) \to$$
$$\to H_0 C_\bullet^X(F) \to H_0 C_\bullet^{X \setminus S}(F) \to 0$$

we get the assertion of Lemma 2.1. The lemma is proved.

2.2. Theorem. Let X be a complex space which is countable at infinity, and S be an arbitrary closed set in X. Then for any analytic sheaf $\mathcal{F}$ on X there exists a spectral sequence, whose initial term occurs in an exact sequence

$$\cdots \to E_{p,q}^2 \to H_p K_\bullet^S(\mathcal{H}_q^X(\mathcal{F})) \to$$
$$\to H_{p-1} \lim^{(1)} K_\bullet(\mathfrak{M}_j;\ \mathcal{H}_{q+1}^X(\mathcal{F})) \to E_{p-1,q}^2 \to \cdots,$$

where $\mathcal{H}_q^X(\mathcal{F})$ is the copresheaf $M \mapsto H_q M(X;\ \mathcal{F})$ on the category of closed sets in X, and whose term E^∞ is the bigraded group associated with a suitable filtration of the graded group $H_\bullet^S(X;\ \mathcal{F})$.

Proof. We choose an arbitrary locally fine resolution

$$0 \to \mathcal{F} \to \mathcal{L}^0 \to \mathcal{L}^1 \to \cdots$$

of the analytic sheaf $\mathscr{F}$ on X. We consider the double complex

$$K_{p,q} = K_p^S(\mathscr{H}_0^X(\mathscr{L}^q)).$$

For its first spectral sequence

$$'E_{p,q}^1 = H_q K_p^S(\mathscr{H}_0^X(\mathscr{L}^{\cdot})).$$

On the other hand, there is an exact sequence

$$0 \to \varprojlim{}^{(1)} K_p(\mathfrak{M}_j; \ \mathscr{H}_{q+1}^X(\mathscr{F})) \to$$
$$\to H_q K_p^S(\mathscr{H}_0^X(\mathscr{L}^{\cdot})) \to K_p^S(\mathscr{H}_q^X(\mathscr{F})) \to 0$$

(cf. Roos [1, p. 3703]). Consequently, the initial term of the first spectral sequence of the double complex $(K_{p,q})$ occurs in the exact sequence

$$\dots \to H_p \varprojlim{}^{(1)} K_{\cdot}(\mathfrak{M}_j; \ \mathscr{H}_{q+1}^X(\mathscr{F})) \to 'E_{p,q}^2 \to H_p K_{\cdot}^S(\mathscr{H}_q^X(\mathscr{F})) \to \dots$$

For the second spectral sequence of the double complex $(K_{p,q})$ by Lemma 2.1 we get

$$"E_{p,q}^1 = 0 \quad \text{for} \quad q \neq 0,$$
$$"E_{p,0}^1 = \Gamma_S(X; \ \mathscr{H}_0(\mathscr{L}^p)).$$

Hence the second spectral sequence degenerates:

$$"E_{p,q}^2 = 0 \quad \text{for} \quad q \neq 0,$$
$$"E_{p,0}^2 = H_p^S(X; \ \mathscr{F}).$$

The theorem is proved.

3. Duality Theorems

3.1. Let X be a complex space which is countable at infinity, S be a closed set in X, $\mathfrak{U}$ be a locally finite covering of the space X by relatively compact open sets, $\mathfrak{B}$ be a locally finite covering of the set $X \setminus S$ by relatively compact open sets, inscribed in $\mathfrak{U}$. For an arbitrary presheaf of Abelian groups F on the category of open sets in X we define the cone $K_S^{\cdot}(F)$ of the chain map

$$C^{\cdot}(\mathfrak{U}; \ F) \to C^{\cdot}(\mathfrak{B}; \ F)$$

(cf. MacLane [1, pp. 67-68]).

3.1.1. Lemma. If $\mathscr{F}$ is a flabby analytic sheaf on X, then

$$H^k K_S^{\cdot}(\mathscr{F}) = 0 \quad \text{for} \quad k \neq 0,$$
$$H^0 K_S^{\cdot}(\mathscr{F}) = \Gamma_S(X; \mathscr{F}).$$

Proof. We make use of the spectral sequence

$$E_2^{p,\,q} = H^p(\mathfrak{U}; \mathscr{H}^q(\mathscr{F})) \Rightarrow H^{p+q}(X; \mathscr{F})$$

(cf. point 1.9.3 of Chapter 1). Since $\mathscr{F}$ is a flabby sheaf, the spectral sequence degenerates. Consequently,

$$H^k(\mathfrak{U}; \mathscr{F}) = 0 \quad \text{for} \quad k \neq 0,$$
$$H^0(\mathfrak{U}; \mathscr{F}) = \Gamma(X; \mathscr{F}).$$

Analogously,

$$H^k(\mathfrak{B}; \mathscr{F}) = 0 \quad \text{for} \quad k \neq 0,$$
$$H^0(\mathfrak{B}; \mathscr{F}) = \Gamma(X \setminus S; \mathscr{F}).$$

Moreover, there is an exact sequence

$$0 \to \Gamma_S(X; \mathscr{F}) \to \Gamma(X; \mathscr{F}) \to \Gamma(X \setminus S; \mathscr{F}) \to 0.$$

From this and the exact sequence of cohomology groups

$$0 \to H^0 K_S^{\cdot}(F) \to H^0(\mathfrak{U}; F) \to H^0(\mathfrak{B}; F) \to \ldots \to H^k K_S^{\cdot}(F) \to$$
$$\to H^k(\mathfrak{U}; F) \to H^k(\mathfrak{B}; F) \to H^{k+1} K_S^{\cdot}(F) \to \ldots$$

we get the assertion of Lemma 3.1.1. The lemma is proved.

3.1.2. Lemma. For an arbitrary analytic sheaf $\mathscr{F}$ on X there exists a spectral sequence

$$E_2^{p,\,q} = H^p K_S^{\cdot}(\mathscr{H}^q(\mathscr{F})) \Rightarrow H_S^{p+q}(X; \mathscr{F}),$$

where $\mathscr{H}^q(\mathscr{F})$ is the presheaf $U \to H^q(U; \mathscr{F})$ on the category of open sets in X.

Proof. Let

$$0 \to \mathscr{F} \to \mathscr{L}^0 \to \mathscr{L}^1 \to \cdots.$$

be a flabby resolution of the sheaf $\mathscr{F}$. We consider the double complex

$$K^{p,\,q} = K_S^p(\mathscr{L}^q).$$

For its first spectral sequence

$$'E_1^{p,\,q} = K_S^p(\mathscr{H}^q(\mathscr{F})),$$
$$'E_2^{p,\,q} = H^p K_S^{\cdot}(\mathscr{H}^q(\mathscr{F})).$$

For the second spectral sequence

$$''E_1^{p,\,q} = 0 \quad \text{for} \quad q \neq 0,$$
$$''E_1^{p,\,0} = \Gamma_S(X;\ \mathscr{L}^p)$$

(cf. Lemma 3.1.1). Consequently, the second spectral sequence degenerates:

$$''E_2^{p,\,q} = 0 \quad \text{for} \quad q \neq 0,$$
$$''E_2^{p,\,0} = H_S^p(X;\ \mathscr{F}).$$

The lemma is proved.

3.1.3. We shall now assume that the coverings $\mathfrak{U}$ and $\mathfrak{V}$ consist of holomorphically complete open sets. Let $\mathscr{F}$ be a coherent analytic sheaf on X. Then by Cartan's theorem (B) the spectral sequence of Lemma 3.1.2 degenerates. We get a natural isomorphism of vector spaces

$$H_S^k(X;\ \mathscr{F}) = H^k K_S^{\cdot}(\mathscr{F}).$$

3.1.4. Each vector space of cochains

$$K_S^k(\mathscr{F}) = C^k(\mathfrak{U};\ \mathscr{F}) \times C^{k-1}(\mathfrak{V};\ \mathscr{F})$$

will be endowed with the product topology (cf. point 2.1.2 of Chapter 3); this is the topology of a Frechet–Schwartz space. We endow the cohomology vector space $H^k K_S(\mathscr{F})$ with the quotient topology. Thus, by means of the isomorphism of point 3.1.3 we define a topology on the cohomology vector space $H_S^k(X;\ \mathscr{F})$. In view of Lemma 3.6 of Chapter 1 this topology is actually independent of the choice of coverings $\mathfrak{U}$ and $\mathfrak{V}$. The associated separated topological vector space $\tilde{H}_S^k(X;\ \mathscr{F})$ (the tilde denotes factorization by the closure of zero) is a Frechet–Schwartz space.

3.2. Let X be a complex space which is countable at infinity, S be a closed set in X, $\mathfrak{U}$ be a locally finite covering of the space X by relatively compact holomorphically complete open sets, $\mathfrak{B}$ be a locally finite covering of the set $X \setminus S$ by relatively compact holomorphically complete open sets, inscribed in $\mathfrak{U}$, and $\mathscr{F}$ be a coherent analytic sheaf on X.

3.2.1. Under the assumptions made, according to Theorem 4.1 of Chapter 3, the spectral sequence of Theorem 1.2 degenerates. We get a natural isomorphism of vector space

$$H_k^c(S;\ \mathscr{F}) = H_k\, K_\bullet^c(\mathscr{H}_0^c(\mathscr{F})).$$

3.2.2. Each vector space of chains

$$K_k^c(\mathscr{H}_0^c(\mathscr{F})) = C_k^c(\mathfrak{U};\ \mathscr{H}_0^c(\mathscr{F})) \times C_{k-1}^c(\mathfrak{B};\ \mathscr{H}_0^c(\mathscr{F}))$$

will be endowed with the topology of locally convex direct sum (cf. point 2.2.2 of Chapter 3), which is the topology of a strong dual to a Frechet–Schwartz space. We endow the homology vector space $H_k K_\bullet^c(\mathscr{H}_0^c(\mathscr{F}))$ with the quotient topology. Finally, by means of the isomorphism of point 3.2.1 we define a topology on the homology vector space $H_k^c(S;\ \mathscr{F})$. Actually, in view of Lemma 3.6 of Chapter 1, this topology is independent of the choice of coverings $\mathfrak{U}$ and $\mathfrak{B}$. The associated separated topological vector space $\widetilde{H}_k^c(S;\ \mathscr{F})$ (the tilde denotes factorization by the closure of zero) is the strong dual to a Frechet–Schwartz space.

3.3. Theorem. Let X be a complex space which is countable at infinity, S be a closed set in X, $\mathscr{F}$ be a coherent analytic sheaf on X. Then the topological vector space $\widetilde{H}_k^c(S;\ \mathscr{F})$ is naturally isomorphic to the strong dual to the topological vector space $\widetilde{H}_S^k(S;\ \mathscr{F})$[3]

$$\{\widetilde{H}_S^k(X;\ \mathscr{F})\}' = \widetilde{H}_k^c(S;\ \mathscr{F}).$$

Proof. Let $\mathfrak{U}$ and $\mathfrak{B}$ be locally finite coverings respectively of the space X and of the open set $X \setminus S$ by relatively compact holomorphically complete open sets, where the covering $\mathfrak{B}$ is inscribed in $\mathfrak{U}$. Then by Theorem 4.1 of Chapter 3, the strong dual to the topological vector space of cochains $K_S^k(\mathscr{F})$ (cf. point 3.1.4) is naturally isomorphic to the topological vector space of chains $K_k^c(\mathscr{H}_0^c(\mathscr{F}))$ (cf. point 3.2.2). The linear map, dual to the coboundary operator

$$\delta:\ K_S^k(\mathscr{F}) \longrightarrow K_S^{k+1}(\mathscr{F}),$$

can be identified naturally with the boundary operator

$$\partial\colon K^c_{k+1}\left(\mathscr{H}^c_0\left(\mathscr{F}\right)\right) \longrightarrow K^c_k\left(\mathscr{H}^c_0\left(\mathscr{F}\right)\right).$$

With the help of Lemma 3.3 of Chapter 1, we get a natural isomorphism of topological vector spaces

$$\left\{\tilde{H}^k K^{\cdot}_S\left(\mathscr{F}\right)\right\}' = \tilde{H}_k K^c_{\cdot}\left(\mathscr{H}^c_0\left(\mathscr{F}\right)\right)$$

(the tilde denotes factorization by the closure of zero). The assertion follows from this by virtue of what was said in points 3.1.4 and 3.2.2. The theorem is proved.

Corollary. The cohomology topological vector space $H_S{}^k(X;\ \mathscr{F})$ is separated if and only if the homology topological vector space $H_{k-1}{}^c(S;\ \mathscr{F})$ is separated.

3.4. Lemma. Let S be a compact set in the domain G of the space $\mathbb{C}^n$, for which there exists a fundamental system of holomorphically complete open neighborhoods, and $\mathscr{F}$ be a coherent analytic sheaf on G, admitting a free resolution of the form

$$0 \longrightarrow 6^{s_n}_G \xrightarrow{\sigma_n} \cdots \xrightarrow{\sigma_2} 6^{s_1}_G \xrightarrow{\sigma_1} 6^{s_0}_G \longrightarrow \mathscr{F} \longrightarrow 0.$$

Then there exists a natural isomorphism of topological vector spaces

$$H_k\left(S;\ \mathscr{F}\right) = H^{n-k} H_n\left(S;\ 6^{s_n}_G\right).$$

Proof. Let $\mathfrak{U}$ and $\mathfrak{V}$ be locally finite coverings respectively of the domain G and the open set $G\setminus S$ by relatively compact holomorphically complete open sets, where the covering $\mathfrak{V}$ is inscribed in $\mathfrak{U}$. We consider the double complex

$$K_{p,\,q} = K^c_p\left(\mathscr{H}^c_0\left(6^{s_{n-q}}_G\right)\right)$$

and the associated single complex

$$K_r = \prod_{p+q=r} K_{p,\,q}.$$

For the first filtration of the complex $K_{\cdot}$ we get

$$'E^1_{p,\,q} = 0 \quad \text{for} \quad q \neq n,$$
$$'E^1_{p,\,n} = K^c_p\left(\mathscr{H}^c_0\left(\mathscr{F}\right)\right).$$

Consequently, the first spectral sequence degenerates:

$$'E^2_{p,\,q}=0 \quad \text{for} \quad q \neq n,$$
$$'E^2_{p,\,n}=H_p(S;\ \mathscr{F}).$$

Thus, the natural continuous linear morphism of complexes

$$'E^1_{p,\,n} \longrightarrow K_{p+n} \tag{1}$$

is an algebraic quasiisomorphism. For the second filtration of the complex K we get

$$''E^1_{p,\,q}=H_q(S;\ \mathcal{O}_G^{s\,n-p}).$$

Consequently, the second spectral sequence also degenerates:

$$''E^2_{p,\,q}=0 \quad \text{for} \quad q \neq n,$$
$$''E^2_{p,\,n}=H^{n-p}H_n(S;\ \mathcal{O}_G^{s\cdot}).$$

We consider the complex

$$L_p = \mathrm{Ker}\ \{K_{n,\,p} \xrightarrow{\partial} K_{n-1,\,p}\}.$$

Then the inclusion of complexes

$$L_p \longrightarrow K_{n,\,p} \tag{2}$$

is an algebraic quasiisomorphism (cf. Godement [1, p. 107]). On the other hand, there exists a natural continuous linear morphism of complexes

$$L_p \longrightarrow ''E^1_{p,\,n}, \tag{3}$$

which is also an algebraic quasiisomorphism. By Lemma 3.6 of Chapter 1 the morphisms of complexes (1)-(3) are topological quasiisomorphisms. Thus they define isomorphisms of topological vector spaces

$$'E^2_{p,\,n} = ''E^2_{p,\,n}.$$

The lemma is proved.

Corollary. Under the assumptions made, the topological vector spaces $H_k(S;\ \mathscr{F})$ are separated.

Proof. Since $\mathrm{inj\,dim}_{\mathcal{O}_G}\mathcal{O}_G = n$, one has

$$H^k_S(G;\; \mathcal{O}_G) = 0 \quad \text{for} \quad k \geq n+1$$

(cf. Corollary 4.3 of Chapter 1). In particular, the coboundary operator

$$\delta \colon K^n_S(\mathcal{O}_G) \longrightarrow K^{n+1}_S(\mathcal{O}_G)$$

has closed image and hence is a homomorphism of Frechet–Schwartz spaces. It follows from this that the boundary operator

$$\partial \colon K^c_{n+1}(\mathcal{H}^c_0(\mathcal{O}_G)) \longrightarrow K^c_n(\mathcal{H}^c_0(\mathcal{O}_G))$$

has closed image, i.e., the topological vector space $H_n(S;\; \mathcal{O}_G)$ is separated. Let U be a relatively compact open neighborhood of the set S in G. By Poincaré duality there exists a natural isomorphism of vector spaces

$$H_n(U;\; \mathcal{O}_G) = \Gamma(U;\; \Omega^n_G)$$

(cf. Corollary 4.2 of Chapter 2). Since the topology on $H_n(U;\; \mathcal{O}_G)$ is compatible with the structure of $\Gamma(U;\; \mathcal{O}_G)$-module, this is an isomorphism of Frechet–Schwartz spaces. Passing to the inductive limit, we get an isomorphism of topological vector spaces, strongly dual to Frechet–Schwartz spaces:

$$H_n(S;\; \mathcal{O}_G) = \Gamma(S;\; \Omega^n_G).$$

Finally, each homomorphism of $\mathcal{O}_G$-modules $\sigma_k\colon \mathcal{O}_G^{s_k} \to \mathcal{O}_G^{s_{k-1}}$ is defined by a holomorphic matrix on G. Hence it defines a continuous linear map

$$\sigma_k\colon \Gamma(S;\; \mathcal{O}^{s_k}_G) \longrightarrow \Gamma(S;\; \mathcal{O}^{s_{k-1}}_G),$$

which is a homomorphism of strong duals to Frechet–Schwartz spaces. The assertion follows directly from Lemma 3.4.

3.5. Theorem. Let X be a complex space which is countable at infinity, and S be a compact set in X, for which there exists a fundamental system of holomorphically complete open neighborhoods. Then for any coherent analytic sheaf $\mathcal{F}$ on X, the topological vector space $H_k(S;\; \mathcal{F})$ is separated and naturally isomorphic to the strong dual to the separated topological vector space $H_S^k(X;\; \mathcal{F})$:

$$\{H^k_S(X;\; \mathcal{F})\}' = H_k(S;\; \mathcal{F}).$$

Proof. We make use of the spectral sequence

$$E^2_{p,\,q} = H^{-p}\,(S;\ \mathscr{H}_q(\mathscr{F})) \Rightarrow H_{p+q}\,(S;\ \mathscr{F})$$

(cf. Theorem 4.1 of Chapter 2). Under the assumptions made, by virtue of Cartan's theorem (B) the spectral sequence degenerates. We get a natural isomorphism of vector spaces

$$H_k(S;\ \mathscr{F}) = \Gamma\,(S:\ \mathscr{H}_k(\mathscr{F})).$$

We represent the set S as a union of a finite number of sufficiently small compact sets S_j ($j = 1, \ldots, m$), for each of which there exists a fundamental system of holomorphically complete open neighborhoods. By the corollary to Lemma 3.4 the topological vector spaces $H_k(S_j;\ \mathscr{F})$ are separated. On the other hand, the natural map,

$$H_k(S;\ \mathscr{F}) \longrightarrow \prod_{1 \leqslant j \leqslant m} H_k(S_j;\ \mathscr{F}),$$

defined by restrictions, is injective and continuous. Consequently, the spaces $H_k(S;\ \mathscr{F})$ are separated and then by the corollary to Theorem 3.3 the spaces $H_S^k(X;\ \mathscr{F})$ are also separated. The theorem is proved.[4]

Corollary. Let X be a holomorphically complete complex space and S be a compact set in X, for which there exists a fundamental system of holomorphically complete open neighborhoods. Then for any coherent analytic sheaf $\mathscr{F}$ on X the topological vector spaces $H^k(X \setminus S;\ \mathscr{F})$ are separated.[5]

Proof. We make use of the cohomology exact sequence connected with the open set $X \setminus S$:

$$\ldots \longrightarrow H^k_S(X;\ \mathscr{F}) \longrightarrow H^k(X;\ \mathscr{F}) \longrightarrow H^k(X \setminus S;\ \mathscr{F}) \longrightarrow H^{k+1}_S(X;\ \mathscr{F}) \longrightarrow \ldots$$

(cf. point 1.8.2 of Chapter 1). From Cartan's theorem (B) we get that for $k \geq 1$ the continuous linear map

$$H^k(X \setminus S;\ \mathscr{F}) \longrightarrow H^{k+1}_S(X;\ \mathscr{F})$$

is bijective. Since by Theorem 3.5 the spaces $H_S^k(X;\ \mathscr{F})$ are separated, the assertion of the corollary is proved.

3.6. Let X be a complex space which is countable at infinity. For each $j = 1, 2, \ldots$ we define by induction a locally finite covering $\mathfrak{M}_j$ of the space X by compact sets so that the following conditions hold:

a) the covering $\mathfrak{M}_j$ consists of sets, small of order $1/j$ with respect to a fixed metric, defining the topology of the space X;

b) the covering $\mathfrak{M}_{j+1}$ is the disjoint union of finite coverings of all sets belonging to the covering $\mathfrak{M}_j$.

For each $j = 1, 2, \ldots$ for an arbitrary presheaf of Abelian groups F on the category of closed sets in X there is defined a natural map of cochain complexes

$$C_c^k (\mathfrak{M}_j;\ F) \longrightarrow C_c^k (\mathfrak{M}_{j+1};\ F)$$

and, consequently, one can define the corresponding inductive limit

$$C_X^k (F) = \varinjlim C_c^k (\mathfrak{M}_j;\ F).$$

For an arbitrary closed set $S \subset X$ we set

$$C_{X \setminus S}^k (\mathfrak{M}_j;\ F) = \prod_{i_0,\ldots,i_k} F (M_{i_0 \ldots i_k}),$$

where the direct sum is taken over those collections of indices $(i_0, \ldots, i_k)$, for which the intersection $M_{i_0 \ldots i_k}$ of the sets $M_{i_0}, \ldots, M_{i_k} \in \mathfrak{M}_j$ is contained in $X \setminus S$. Analogously to the preceding, there are defined natural maps of cochain complexes

$$C_{X \setminus S}^k (\mathfrak{M}_j;\ F) \longrightarrow C_{X \setminus S}^k (\mathfrak{M}_{j+1};\ F)$$

and, consequently, one can define the corresponding inductive limit

$$C_{X \setminus S}^k (F) = \varinjlim C_{X \setminus S}^k (\mathfrak{M}_j;\ F).$$

We define the cone $K_c^\cdot (F)$ of the cochain map

$$C_{X \setminus S}^\cdot (F) \longrightarrow C_X^\cdot (F),$$

by defining its components and differentials in the familiar way (cf. MacLane [1, pp. 67–68]).

3.6.1. Lemma. If $\mathscr{F}$ is a flabby analytic sheaf on X, then

$$H^k K_c^\cdot (\mathscr{F}) = 0 \quad \text{for} \quad k \neq 0;$$
$$H^0 K_c^\cdot (\mathscr{F}) = \Gamma_c (S;\ \mathscr{F}).$$

Proof. Since the sheaf $\mathscr{F}$ is flabby, one has

$$H_c^k(\mathfrak{M}_j;\ \mathscr{F}) = 0 \quad \text{for} \quad k \neq 0,$$
$$H_c^0(\mathfrak{M}_j;\ \mathscr{F}) = \Gamma_c(X;\ \mathscr{F})$$

for each $j = 1, 2, \ldots$ (cf. Godement [1, p. 234]). Passing to the inductive limit, we get

$$H^k C_X^{\cdot}(\mathscr{F}) = 0 \quad \text{for} \quad k \neq 0,$$
$$H^0 C_X^{\cdot}(\mathscr{F}) = \Gamma_c(X;\ \mathscr{F}).$$

Analogously,

$$H^k C_{X \setminus S}^{\cdot}(\mathscr{F}) = 0 \quad \text{for} \quad k \neq 0,$$
$$H^0 C_{X \setminus S}^{\cdot}(\mathscr{F}) = \Gamma_c(X \setminus S;\ \mathscr{F}).$$

Moreover, there is an exact sequence

$$0 \longrightarrow \Gamma_c(X \setminus S;\ \mathscr{F}) \longrightarrow \Gamma_c(X;\ \mathscr{F}) \longrightarrow \Gamma_c(S;\ \mathscr{F}) \longrightarrow 0.$$

From this and the exact sequence of cohomology groups

$$0 \longrightarrow H^0 C_{X \setminus S}^{\cdot}(F) \longrightarrow H^0 C_X^{\cdot}(F) \longrightarrow H^v K_c^{\cdot}(F) \longrightarrow \cdots$$
$$\cdots \longrightarrow H^k C_{X \setminus S}^{\cdot}(F) \longrightarrow H^k C_X^{\cdot}(F) \longrightarrow H^k K_c^{\cdot}(\mathscr{F}) \longrightarrow H^{k+1} C_{X \setminus S}^{\cdot}(F) \longrightarrow \cdots$$

we get the assertion of Lemma 3.6.1. The lemma is proved.

3.6.2. Lemma. For an arbitrary analytic sheaf $\mathscr{F}$ on X there exists a spectral sequence

$$E_2^{p,\,q} = H^p K_c^{\cdot}(\mathscr{H}^q(\mathscr{F})) \Rightarrow H_c^{p+q}(S;\ \mathscr{F}),$$

where $\mathscr{H}^q(\mathscr{F})$ is the presheaf $M \mapsto H^q(M;\ \mathscr{F})$ on the category of closed sets in X.

Proof. We choose an arbitrary flabby resolution

$$0 \longrightarrow \mathscr{F} \longrightarrow \mathscr{L}^0 \longrightarrow \mathscr{L}^1 \longrightarrow \cdots$$

of the analytic sheaf $\mathscr{F}$ on X. We consider the double complex

$$K^{p,\,q} = K_c^p(\mathscr{L}^q).$$

For its first spectral sequence

$$'E_1^{p,\,q} = K_c^p\,(\mathcal{H}^q\,(\mathcal{F}));$$
$$'E_2^{p,\,q} = H^p\,K_c^{\cdot}\,(\mathcal{H}^q\,(\mathcal{F})).$$

For the second spectral sequence

$$''E_1^{p,\,q} = 0 \quad \text{for} \quad q \neq 0,$$
$$''E_1^{p,\,0} = \Gamma_c\,(S;\; \mathcal{L}^p)$$

(cf. Lemma 3.6.1). Consequently, the second spectral sequence degenerates:

$$''E_2^{p,\,q} = 0 \quad \text{for} \quad q \neq 0,$$
$$''E_2^{p,\,0} = H_c^p\,(S;\; \mathcal{F}).$$

The lemma is proved.

3.6.3. Now we shall assume that the coverings $\mathfrak{M}_j$ $(j = 1, 2, \ldots)$ consist of compact sets, for each of which there exists a fundamental system of holomorphically complete open neighborhoods. Let $\mathcal{F}$ be a coherent analytic sheaf on X. Then by Cartan's theorem (B) the spectral sequence of Lemma 3.6.2 degenerates. We get a natural isomorphism of vector spaces

$$H_c^k\,(S;\; \mathcal{F}) = H^k\,K_c^{\cdot}\,(\mathcal{F}).$$

3.6.4. We endow the vector spaces of cochains $C_X{}^k(\mathcal{F})$ and $C_{X\backslash S}{}^{k+1}(\mathcal{F})$ with the topologies of locally convex inductive limits. Since they are strict inductive limits, the spaces $C_X{}^k(\mathcal{F})$ and $C_{X\backslash S}{}^{k+1}(\mathcal{F})$ are strong duals to Frechet–Schwartz spaces (cf. point 3.1.2 of Chapter 3). We endow each vector space of cochains

$$K_c^k\,(\mathcal{F}) = C_X^k\,(\mathcal{F}) \times C_{X\backslash S}^{k+1}\,(\mathcal{F})$$

with the topology of a direct sum, which is also the topology of a strong dual to a Frechet–Schwartz space. We endow the cohomology vector space $H^k K_c^{\cdot}(\mathcal{F})$ with the quotient topology. Thus, by means of the isomorphism of point 3.6.3 we define a locally convex topology on the cohomology vector space $H_c{}^k(S;\; \mathcal{F})$. In view of Lemma 3.6 of Chapter 1 this topology is actually independent of the choice of coverings $\mathfrak{M}_j$ $(j = 1, 2, \ldots)$. The associated separated topological vector space $\tilde{H}_c{}^k(S;\; \mathcal{F})$ (the tilde denotes factorization by the closure of zero) is the strong dual to a Frechet–Schwartz space.

3.7. Under the assumptions of point 3.6 let the coverings $\mathfrak{M}_j$ $(j = 1,$ $2, \ldots)$ be sufficiently fine and consist of compact sets, for each of which there exists a fundamental system of holomorphically complete open neighborhoods. Let $\mathscr{F}$ be a coherent analytic sheaf on X.

3.7.1. Lemma. There exists a natural isomorphism of vector spaces

$$H_k^S(X;\ \mathscr{F}) = H_k\, K_{\bullet}^S\, (\mathscr{H}_0^X\, (\mathscr{F})).$$

Proof. Let M be a sufficiently small compact set in X, for which there exists a fundamental system of holomorphically complete open neighborhoods. One can assume that an open neighborhood of the set M is realized as an analytic set in the domain G of the space $\mathbf{C}^n$. Then there is a natural isomorphism of vector spaces

$$H_q^M(X;\ \mathscr{F}) = \operatorname{Ext}_{\mathscr{O}_G,\ M}^{n-q}\, (G;\ \mathscr{F}^G,\ \Omega_G^n)$$

(cf. Corollary 3.4 of Chapter 2). On the other hand, by Cartan's theorem (B),

$$H^q(M;\ \mathscr{F}) = 0 \quad \text{for} \quad q \neq 0,$$

consequently,

$$\operatorname{Ext}_{\mathscr{O}_G,\ M}^{n-q}\, (G;\ \mathscr{F}^G,\ \Omega_G^n) = 0 \quad \text{for} \quad q \neq 0$$

(cf. Corollary 3.12.2 of Chapter 1). In particular, for the sets $M_{i_0}, \ldots, M_{i_p} \in \mathfrak{M}_j$,

$$H_q^{M_{i_0}\cdots i_p}(X;\ \mathscr{F}) = 0 \quad \text{for} \quad q \neq 0.$$

Thus, from the exact sequence of Theorem 2.2 we get

$$\begin{aligned}
E_{p,\ q}^2 &= 0 \quad \text{for} \quad q \neq 0,\\
E_{p,\ 0}^2 &= H_p K_{\bullet}^S\, (\mathscr{H}_0^X\, (\mathscr{F})).
\end{aligned}$$

In other words, the spectral sequence of Theorem 2.2 degenerates. The lemma is proved.

3.7.2. The vector spaces of chains $C_k^X(\mathscr{H}_0^X(\mathscr{F}))$ and $C_{X\setminus S}^{k+1} \times$ $(\mathscr{H}_0^X(\mathscr{F}))$ will be endowed with the topologies of projective limits (cf. point 2.1), which are topologies of Frechet–Schwartz spaces (cf. point 3.2.2 of Chapter 3). We endow each vector space of chains

$$K_k^S \left(\mathcal{H}_0^X \left(\mathcal{F} \right) \right) = C_k^X \left(\mathcal{H}_0^X \left(\mathcal{F} \right) \right) \times C_{k+1}^{X \setminus S} \left(\mathcal{H}_0^X \left(\mathcal{F} \right) \right)$$

with the product topology, which is also the topology of a Frechet–Schwartz space. We endow the homology vector space $H_k K^S_{\cdot} (\mathcal{H}_0^X(\mathcal{F}))$ with the quotient topology. Finally, by means of the isomorphism of Lemma 3.7.1 we define a locally convex topology on the homology vector space $H_k^S(X; \mathcal{F})$. In view of Lemma 3.6 of Chapter 1, this topology is actually independent of the choice of the coverings $\mathfrak{M}_j$ ($j = 1, 2, \ldots$). The associated separated topological vector space $\tilde{H}_k^S(X; \mathcal{F})$ (the tilde denotes factorization by the closure of zero) is a Frechet–Schwartz space.

3.8. Theorem. Let X be a complex space which is countable at infinity, S be a closed set in X, $\mathcal{F}$ be a coherent analytic sheaf on X. Then the topological vector space $\tilde{H}_k^S(X; \mathcal{F})$ is naturally isomorphic to the strong dual to the topological vector space $\tilde{H}_c^k(S; \mathcal{F})$[6]:

$$\{\tilde{H}_c^k (S; \mathcal{F})\}' = \tilde{H}_k^S (X; \mathcal{F}).$$

Proof. Let $(\mathfrak{M}_j)$ be a sequence of coverings satisfying the conditions of points 3.6 and 3.7. For any sets $M_{i_0}, \ldots, M_{i_k} \in \mathfrak{M}_j$ the strong dual to the topological vector space $\Gamma(M_{i_0 \ldots i_k}; \mathcal{F})$ is naturally isomorphic to the topological vector space $H_0^{M_{i_0 \ldots i_k}}(X; \mathcal{F})$ (cf. Corollary 3.12.3 of Chapter 1). The strong dual to the topological vector space $C_X^k(\mathcal{F})$ (cf. point 3.6) is naturally isomorphic to the topological vector space of chains $C_k^X(\mathcal{H}_0^X \times (\mathcal{F}))$ (cf. point 2.1). Consequently, the strong dual to the topological vector space of cochains $K_c^k(\mathcal{F})$ (cf. point 3.6.4) is naturally isomorphic to the topological vector space of chains $K_k^S(\mathcal{H}_0^X(\mathcal{F}))$ (cf. point 3.7.2). The linear map, dual to the coboundary operator

$$\delta \colon K_c^k (\mathcal{F}) \to K_c^{k+1} (\mathcal{F}),$$

can be identified naturally with the boundary operator

$$\partial \colon K_{k+1}^S (\mathcal{H}_0^X (\mathcal{F})) \to K_k^S (\mathcal{H}_0^X (\mathcal{F})).$$

With the help of Lemma 3.3 of Chapter 1 we get a natural isomorphism of topological vector spaces

$$\{\tilde{H}^k K_c^{\cdot} (\mathcal{F})\}' = \tilde{H}_k K_{\cdot}^S (\mathcal{H}_0^X(\mathcal{F}))$$

(the tilde denotes factorization by the closure of zero). The assertion of Theorem 3.8 follows from this in view of what was said in points 3.6.4 and 3.7.2. The theorem is proved.

Corollary. The cohomology topological vector space $H_c^k(S; \mathscr{F})$ is separated if and only if the homology topological vector space $H_{k-1}^S(X; \mathscr{F})$ is separated.

3.9. Theorem. Let X be a complex space which is countable at infinity, and S be a compact set in X, for which there exists a fundamental system of holomorphically complete open neighborhoods. Then for any coherent analytic sheaf $\mathscr{F}$ on X

$$H_k^S(X; \mathscr{F}) = 0 \quad \text{for} \quad k \neq 0,$$

and the topological vector space $H_0^S(X; \mathscr{F})$ is separated and naturally isomorphic to the strong dual to the topological vector space $\Gamma(S; \mathscr{F})$:

$$H_0^S(X; \mathscr{F}) = \{\Gamma(S; \mathscr{F})\}'.$$

Proof. By Cartan's theorem (B),

$$H^k(S; \mathscr{F}) = 0 \quad \text{for} \quad k \neq 0.$$

Consequently, the topological vector spaces $H_k^S(X; \mathscr{F})$ are separated for all k (cf. the corollary to Theorem 3.8). To finish the proof it suffices to apply Theorem 3.8. The theorem is proved.[7]

4. Inductive and Projective Limits

4.1. Let X be a complex space which is countable at infinity and

$$X_1 \subset X_2 \subset \ldots \subset X_j \subset \ldots$$

be an increasing sequence of open sets of it, whose union coincides with X:

$$\bigcup_{j=1}^{\infty} X_j = X.$$

Then for an arbitrary coherent analytic sheaf $\mathscr{F}$ on X the homology topological vector spaces $H_k^c(X_j; \mathscr{F})$ ($j = 1, 2, \ldots$) form an inductive system.

For the locally convex inductive limit of this system, there is defined a natural continuous linear map

$$\lim_{\longrightarrow} H_k^c(X_j;\ \mathscr{F}) \to H_k^c(X;\ \mathscr{F}). \tag{1}$$

Theorem. The map (1) is an isomorphism of topological vector spaces.

Proof. For each $j = 1, 2, \ldots$ we choose a countable covering $\mathfrak{U}_j$ of the open set X_j by holomorphically complete open sets. Then the space of chains $C_k{}^c(\mathfrak{U}_j;\ \mathscr{H}_0{}^c(\mathscr{F}))$ is the strong dual to a Frechet–Schwartz space and there is a natural isomorphism of topological vector spaces

$$H_k^c(X_j;\ \mathscr{F}) = H_k^c(\mathfrak{U}_j;\ \mathscr{H}_0^c(\mathscr{F}))$$

(cf. Lemma 2.2.1 of Chapter 3). We shall assume that $\mathfrak{U}_j \subset \mathfrak{U}_{j+1}$ for each $j = 1, 2, \ldots$, and we set $\mathfrak{U} = \cup \mathfrak{U}_j$. Then the space of chains $C_k{}^c(\mathfrak{U};\ \mathscr{H}_0{}^c(\mathscr{F}))$ is the strong dual to a Frechet–Schwartz space and there is a natural isomorphism of topological vector spaces

$$H_k^c(X;\ \mathscr{F}) = H_k^c(\mathfrak{U};\ \mathscr{H}_0^c(\mathscr{F}))$$

(cf. Lemma 2.2.1 of Chapter 3). Moreover, $C_k{}^c(\mathfrak{U};\ \mathscr{H}_0{}^c(\mathscr{F}))$ is the strict locally convex inductive limit of the spaces $C_k{}^c(\mathfrak{U}_j;\ \mathscr{H}_0{}^c(\mathscr{F}))$ $(j = 1, 2, \ldots)$. It follows directly from this that the map (1), which is obviously an isomorphism of vector spaces, is also an isomorphism of topological vector spaces. The theorem is proved.

4.1.1. The cohomology topological vector spaces $H^k(X_j;\ \mathscr{F})$ $(j = 1, 2, \ldots)$ form a projective system. There is defined a natural continuous linear map

$$H^k(X;\ \mathscr{F}) \to \lim_{\longleftarrow} H^k(X_j;\ \mathscr{F}) \tag{2}$$

and a natural continuous linear map of separated spaces

$$\tilde{H}^k(X;\ \mathscr{F}) \to \lim_{\longleftarrow} \tilde{H}^k(X_j;\ \mathscr{F}). \tag{3}$$

Moreover, by Theorem 2.3 of Chapter 3 the map (3) can be naturally identified with the dual to the map (1). Thus the following theorem is proved.

Theorem. The map (3) is an isomorphism of topological vector spaces.

4.1.2. Corollary. If the spaces $H^k(X; \mathscr{F})$ and $H^k(X_j; \mathscr{F})$ $(j = 1, 2, ...)$ are separated, then the map (2) is an isomorphism of topological vector spaces.[8]

4.2. Theorem. Let the topological vector spaces $H_k{}^c(X_j; \mathscr{F})$ $(j = 1, 2, ...)$ be separated. Then the space $H_k{}^c(X; \mathscr{F})$ is separated if and only if the following condition holds: for each $j = 1, 2, ...$ and each bounded set B in $H_k{}^c(X_j; \mathscr{F})$, whose image in the space $H_k{}^c(X; \mathscr{F})$ is equal to zero, there exists a $j' > j$ such that the image of the set B in the space $H_k{}^c(X_{j'}; \mathscr{F})$ is equal to zero.

Proof. Let $\mathfrak{U}_j$ for each $j = 1, 2, ...$ be a countable covering of the set X_j by holomorphically complete open sets, while $\mathfrak{U}_j \subset \mathfrak{U}_{j+1}$. For $\mathfrak{U} = \bigcup \mathfrak{U}_j$ the space of chains $C_k{}^c(\mathfrak{U}; \mathscr{H}_0{}^c(\mathscr{F}))$ can be naturally identified with the strict locally convex inductive limit of the spaces $C_k{}^c(\mathfrak{U}_j; \mathscr{H}_0{}^c(\mathscr{F}))$ and is the strong dual to a Frechet–Schwartz space.

Let us assume that the condition of the theorem holds. We show that then the space $H_k{}^c(X; \mathscr{F})$ is separated, i.e., that the subspace $\partial C_{k+1}{}^c(\mathfrak{U}; \mathscr{H}_0{}^c(\mathscr{F}))$ in $C_k{}^c(\mathfrak{U}; \mathscr{H}_0{}^c(\mathscr{F}))$ is closed (cf. Lemma 2.2.1 of Chapter 3). By the Banach–Dieudonné theorem (cf. Bourbaki [2, p. 224]) for this, it suffices that the intersection $B \cap \partial C_{k+1}{}^c(\mathfrak{U}; \mathscr{H}_0{}^c(\mathscr{F}))$ be closed in $C_k{}^c(\mathfrak{U}; \mathscr{H}_0{}^c(\mathscr{F}))$ for each closed bounded set B in $C_k{}^c(\mathfrak{U}; \mathscr{H}_0{}^c(\mathscr{F}))$. On the other hand, for some integer $j > 0$ the set B is contained in $C_k{}^c(\mathfrak{U}_j; \mathscr{H}_0{}^c(\mathscr{F}))$. Since the image of the intersection $B \cap \partial C_{k+1}{}^c(\mathfrak{U}; \mathscr{H}_0{}^c(\mathscr{F}))$ in the space $H_k{}^c(\mathfrak{U}; \mathscr{H}_0{}^c(\mathscr{F}))$ is equal to zero, by the condition of the theorem there exists a $j' > j$, such that the image of the intersection $B \cap \partial C_{k+1}{}^c(\mathfrak{U}; \mathscr{H}_0{}^c(\mathscr{F}))$ in the space $H_k{}^c(\mathfrak{U}_{j'}; \mathscr{H}_0{}^c(\mathscr{F}))$ is equal to zero. The latter means that

$$B \cap \partial C_{k+1}^c(\mathfrak{U}; \mathscr{H}_0^c(\mathscr{F})) \subset \partial C_{k+1}^c(\mathfrak{U}_{j'}; \mathscr{H}_0^c(\mathscr{F})),$$

i.e., the intersection $B \cap \partial C_{k+1}{}^c(\mathfrak{U}; \mathscr{H}_0{}^c(\mathscr{F}))$ coincides with the intersection $B \cap \partial C_{k+1}{}^c(\mathfrak{U}_{j'}; \mathscr{H}_0{}^c(\mathscr{F}))$. By hypothesis the space $H_k{}^c(\mathfrak{U}_{j'}; \mathscr{H}_0{}^c(\mathscr{F}))$ is separated, i.e., the subspace $\partial C_{k+1}{}^c(\mathfrak{U}_{j'}; \mathscr{H}_0{}^c(\mathscr{F}))$ is closed in $C_k{}^c(\mathfrak{U}_{j'}; \mathscr{H}_0{}^c(\mathscr{F}))$. The assertion is proved.

We show that conversely, too, the condition of the theorem holds if the space $H_k{}^c(X; \mathscr{F})$ is separated. Let B be a bounded set in $H_k{}^c(X_j; \mathscr{F})$, whose image in the space $H_k{}^c(X; \mathscr{F})$ is equal to zero. One can assume that there is given a bounded set B in $Z_k{}^c(\mathfrak{U}_j; \mathscr{H}_0{}^c(\mathscr{F})) = \text{Ker } \{C_k{}^c(\mathfrak{U}_j;$ $\mathscr{H}_0{}^c(\mathscr{F})) \xrightarrow{\partial} C_{k-1}{}^c(\mathfrak{U}_j; \mathscr{H}_0{}^c(\mathscr{F}))\}$, whose image in the space $H_k{}^c(\mathfrak{U};$ $\mathscr{H}_0{}^c(\mathscr{F}))$ is equal to zero. The latter means that $B \subset \partial C_{k+1}{}^c(\mathfrak{U}; \mathscr{H}_0{}^c(\mathscr{F}))$. Since the space $H_k{}^c(\mathfrak{U}; \mathscr{H}_0{}^c(\mathscr{F}))$ is separated, the subspace $\partial C_{k+1}{}^c(\mathfrak{U};$ $\mathscr{H}_0{}^c(\mathscr{F}))$ is closed in $C_k{}^c(\mathfrak{U}; \mathscr{H}_0{}^c(\mathscr{F}))$. Consequently, $\partial C_{k+1}{}^c(\mathfrak{U};$ $\mathscr{H}_0{}^c(\mathscr{F}))$ is the strong dual to a Frechet–Schwartz space and hence can be identified naturally with the locally convex inductive limit of the spaces $\partial C_{k+1}{}^c(\mathfrak{U}_j; \mathscr{H}_0{}^c(\mathscr{F}))$ $(j = 1, 2, \ldots)$. Hence there exists a $j' > j$ such that the set B is contained in $\partial C_{k+1}{}^c(\mathfrak{U}_{j'}; \mathscr{H}_0{}^c(\mathscr{F}))$. This means that the image of the set B in the space $H_k{}^c(\mathfrak{U}_{j'}; \mathscr{H}_0{}^c(\mathscr{F}))$ is equal to zero. The theorem is proved.

4.2.1. Corollary. Let the topological vector spaces $H_k{}^c(X_j; \mathscr{F})$ $(j = 1, 2, \ldots)$ be separated. Then the space $H_k{}^c(X; \mathscr{F})$ is separated if the following condition holds: for each $j = 1, 2, \ldots$ the kernel of the map

$$H_k^c(X_j; \mathscr{F}) \to H_k^c(X; \mathscr{F})$$

is contained in the kernel of the map

$$H_k^c(X_j; \mathscr{F}) \to H_k^c(X_{j+1}; \mathscr{F}).$$

4.2.2. Corollary. Let the spaces $H_k{}^c(X_j; \mathscr{F})$ $(j = 1, 2, \ldots)$ be separated and suppose that for each $j = 1, 2, \ldots$ the natural map $H_k{}^c(X_j; \mathscr{F}) \to H_k{}^c(X_{j+1}; \mathscr{F})$ is injective. Then the space $H_k{}^c(X; \mathscr{F})$ is separated.

4.2.3. Theorem. Let the topological vector spaces $H^k(X_j; \mathscr{F})$ $(j = 1, 2, \ldots)$ be separated. Then the space $H^k(X; \mathscr{F})$ is separated if and only if the following condition holds: for each $j = 1, 2, \ldots$ and each neighborhood V of zero in the space $H^{k-1}(X_j; \mathscr{F})$ there exists a $j' > j$, such that the image

$$\text{Im } \{H^{k-1}(X_{j'}; \mathscr{F}) \to H^{k-1}(X_j; \mathscr{F})\}$$

is contained in the sum

$$V + \text{Im } \{H^{k-1}(X; \mathscr{F}) \to H^{k-1}(X_j; \mathscr{F})\}.$$

The map (2) is an isomorphism of topological vector spaces if and only if this is the case.[9]

Proof. Under the assumptions made the spaces $H_{k-1}{}^c(X_j; \mathscr{F})$ ($j = 1, 2, \ldots$) are separated (by the corollary to Theorem 2.3 of Chapter 3). Moreover, it is obvious that the polar of the set Im $\{H^{k-1}(X_j; \mathscr{F}) \to H^{k-1}(X_j; \mathscr{F})\}$ in $H_{k-1}{}^c(X_j; \mathscr{F})$ coincides with the kernel Ker $\{H_{k-1}{}^c(X_j; \mathscr{F}) \to H_{k-1}{}^c(X_j; \mathscr{F})\}$. According to Theorem 4.2 the condition of Theorem 4.2.3 is equivalent with the separatedness of the space $H_{k-1}{}^c(X; \mathscr{F})$. In view of the corollary to Theorem 2.3 of Chapter 3, Theorem 4.2.3 is proved.[10]

4.2.4. Corollary. Let the topological vector spaces $H^k(X_j; \mathscr{F})$ ($j = 1, 2, \ldots$) be separated. Then the space $H^k(X; \mathscr{F})$ is separated [and the map (2) is an isomorphism], if the following condition holds: for each $j = 1, 2, \ldots$ the image of the map

$$H^{k-1}(X; \mathscr{F}) \to H^{k-1}(X_j; \mathscr{F})$$

is everywhere dense in the image of the map[11]

$$H^{k-1}(X_{j+1}; \mathscr{F}) \to H^{k-1}(X_j; \mathscr{F}).$$

4.2.5. Theorem. Let the spaces $H^k(X_j; \mathscr{F})$ ($j = 1, 2, \ldots$) be separated and suppose for each $j = 1, 2, \ldots$ the image of the space $H^{k-1}(X_{j+1}; \mathscr{F})$ is everywhere dense in $H^{k-1}(X_j; \mathscr{F})$. Then the space $H^k(X; \mathscr{F})$ is separated and consequently the map (2) is an isomorphism of topological vector spaces.[12]

Proof. Under the assumptions made the spaces $H_{k-1}{}^c(X_j; \mathscr{F})$ ($j = 1, 2, \ldots$) are separated and the maps $H_{k-1}{}^c(X_j; \mathscr{F}) \to H_{k-1}{}^c(X_{j+1}; \mathscr{F})$ are injective. By Corollary 4.2.2 the space $H_{k-1}{}^c(X; \mathscr{F})$ is separated, and hence the space $H^k(X; \mathscr{F})$ is also separated. The theorem is proved.

4.3. We consider the cohomology topological vector spaces $H_c{}^k(X_j; \mathscr{F})$ ($j = 1, 2, \ldots$). They form an inductive system and for the locally convex inductive limit of this system there is defined a natural continuous linear map

$$\varinjlim H_c^k(X_j; \mathscr{F}) \to H_c^k(X; \mathscr{F}). \tag{4}$$

Theorem. The map (4) is an isomorphism of topological vector spaces.

Proof. We give a sequence of coverings $\mathfrak{M}_i$ $(i = 1, 2, \ldots)$ of the space X, satisfying the conditions of points 3.6 and 3.7. We define the cochain spaces $C_X{}^k(\mathscr{F})$ and $C_{X_j}{}^k(\mathscr{F})$ $(j = 1, 2, \ldots)$ (cf. point 3.6). They are strong duals to Frechet–Schwartz spaces. Here the topology of the space $C_X{}^k(\mathscr{F})$ is the strongest locally convex topology for which the maps $C_{X_j}{}^k(\mathscr{F}) \to C_X{}^k(\mathscr{F})$ $(j = 1, 2, \ldots)$ are continuous. Since there are natural isomorphisms of topological vector spaces

$$H_c^k(U; \mathscr{F}) = H^k C_U^{\cdot}(\mathscr{F})$$

(U is an arbitrary open set in X), we get from this that the map (4) is an isomorphism of topological vector spaces, since it is obviously an isomorphism of vector spaces (without topology). The theorem is proved.

4.3.1. The topological vector spaces $H_k(X_j; \mathscr{F})$ $(j = 1, 2, \ldots)$ form a projective system. There is defined a natural continuous linear map

$$H_k(X; \mathscr{F}) \to \varprojlim H_k(X_j; \mathscr{F}) \tag{5}$$

and a natural continuous linear map of separated spaces

$$\tilde{H}_k(X; \mathscr{F}) \to \varprojlim \tilde{H}_k(X_j; \mathscr{F}). \tag{6}$$

Moreover, by Theorem 3.3 of Chapter 3, the map (6) can be identified naturally with the dual of (4). This proves the following

Theorem. The map (6) is an isomorphism of topological vector spaces.

4.3.2. Corollary. If the spaces $H_k(X; \mathscr{F})$ and $H_k(X_j; \mathscr{F})$ $(j = 1, 2, \ldots)$ are separated, then the map (5) is an isomorphism of topological vector spaces.

4.4. Theorem. Let the topological vector spaces $H_c^k(X_j; \mathscr{F})$ $(j = 1, 2, \ldots)$ be separated. Then the space $H_c^k(X; \mathscr{F})$ is separated if and only if the following condition holds: for each $j = 1, 2, \ldots$ and each bounded set B in $H_c^k(X_j; \mathscr{F})$ whose image in $H_c^k(X; \mathscr{F})$ is equal to zero, there exists a $j' > j$ such that the image of the set B in the space $H_c^k(X_{j'}; \mathscr{F})$ is equal to zero.

Proof. We define, as above (cf. the proof of Theorem 4.3), the cochain spaces $C_X{}^k(\mathscr{F})$ and $C_{X_j}{}^k(\mathscr{F})$ $(j = 1, 2, \ldots)$, which are strong duals of Frechet–Schwartz spaces. Here the space $C_X{}^k(\mathcal{F})$ can be identified naturally with the locally convex limit of the spaces $C_{X_j}{}^k(\mathcal{F})$ $(j = 1, 2, \ldots)$.

Let us assume that the condition of the theorem holds. We show that then the space $H_c{}^k(X; \mathcal{F})$ is separated, i.e., that the subspace $\delta C_X{}^{k-1}(\mathcal{F})$ in $C_X{}^k(\mathcal{F})$ is closed. By the Banach–Dieudonné theorem (cf. Bourbaki [2, p. 224]), for this it suffices that the intersection $B \cap \delta C_X{}^{k-1}(\mathcal{F})$ be closed in $C_X{}^k(\mathcal{F})$ for each closed bounded set B in $C_X{}^k(\mathcal{F})$. On the other hand, for some integer $j > 0$ the set B is contained in $C_{X_j}{}^k(\mathcal{F})$. Since the image of the intersection $B \cap \delta C_X{}^{k-1}(\mathcal{F})$ in the space $H^k C'_X(\mathcal{F})$ is equal to zero, by hypothesis there exists a $j' > j$ such that the image of the intersection $B \cap \delta C_X{}^{k-1}(\mathcal{F})$ in the space $H^k C'_{X_{j'}}(\mathcal{F})$ is equal to zero. The latter means that

$$B \cap \delta C_X^{k-1}(\mathcal{F}) \subset \delta C_{X_{j'}}^{k-1}(\mathcal{F}),$$

i.e., the intersection $B \cap \delta C_X{}^{k-1}(\mathcal{F})$ coincides with the intersection $B \cap \delta C_{X_{j'}}{}^{k-1}(\mathcal{F})$. By hypothesis the space $H^k C'_{X_{j'}}(\mathcal{F})$ is separated, i.e., the subspace $\delta C_{X_{j'}}{}^{k-1}(\mathcal{F})$ is closed in $\delta C_{X_{j'}}{}^k(\mathcal{F})$. Consequently, the intersection $B \cap \delta C_{X_{j'}}{}^{k-1}(\mathcal{F})$ is compact in $C_{X_{j'}}{}^k(\mathcal{F})$, and hence closed in $C_X{}^k(\mathcal{F})$. The assertion is proved.

We show that conversely, too, the condition of the theorem holds if the space $H_c{}^k(X; \mathcal{F})$ is separated. Let B be a bounded set in $H_c{}^k(X_j; \mathcal{F})$, whose image in the space $H_c{}^k(X; \mathcal{F})$ is equal to zero. One can assume that there is given a bounded set B in $Z_{X_j}{}^k(\mathcal{F}) = \mathrm{Ker}\,\{C_{X_j}{}^k(\mathcal{F}) \xrightarrow{\partial} C_{X_j}{}^{k+1}\times(\mathcal{F})\}$, whose image in $H^k C'_X(\mathcal{F})$ is equal to zero. The latter means that $B \subset \delta C_X{}^{k-1}(\mathcal{F})$. Since the space $H^k C'_X(\mathcal{F})$ is separated, the subspace $\delta C_X{}^{k-1}\times(\mathcal{F})$ is closed in $C_X{}^k(\mathcal{F})$. Consequently, $\delta C_X{}^{k-1}(\mathcal{F})$ is the strong dual to a Frechet–Schwartz space and hence can be identified naturally with the locally convex inductive limit of the spaces $\delta C_{X_j}{}^{k-1}(\mathcal{F})$ $(j = 1, 2, \ldots)$. Hence there exists a $j' > j$, such that the set B is contained in $\delta C_{X_{j'}}{}^{k-1}(\mathcal{F})$. This means that the image of the set B in the space $H^k C'_{X_{j'}}(\mathcal{F})$ is equal to zero. The theorem is proved.

4.4.1. Corollary. Let the topological vector spaces $H_c{}^k(X_j; \mathcal{F})$ $(j = 1, 2, \ldots)$ be separated. Then the space $H_c{}^k(X; \mathcal{F})$ is separated if the following condition holds: for each $j = 1, 2, \ldots$ the kernel of the map

$$H^k_c(X_j; \; \mathscr{F}) \to H^k_c(X; \; \mathscr{F})$$

is contained in the kernel of the map

$$H^k_c(X_j; \; \mathscr{F}) \to H^k_c(X_{j+1}; \; \mathscr{F}).$$

4.4.2. Corollary. Let the spaces $H_c{}^k(X_j; \; \mathscr{F})$ $(j' = 1, 2, \ldots)$ be separated and suppose for each $j = 1, 2, \ldots$ the natural map $H_c{}^k(X_j; \; \mathscr{F}) \to H_c{}^k(X_{j+1}; \; \mathscr{F})$ is injective. Then the space $H_c{}^k(X; \; \mathscr{F})$ is separated.

4.4.3. Theorem. Let the topological vector spaces $H_k(X_j; \; \mathscr{F})$ $(j = 1, 2, \ldots)$ be separated. Then the space $H_k(X; \; \mathscr{F})$ is separated if and only if the following condition holds: for each $j = 1, 2, \ldots$ and each neighborhood V of zero in the space $H_{k+1}(X_j; \; \mathscr{F})$ there exists a $j' > j$ such that the image

$$\mathrm{Im}\, \{H_{k+1}(X_{j'}; \; \mathscr{F}) \to H_{k+1}(X_j; \; \mathscr{F})\}$$

is contained in the sum

$$V + \mathrm{Im}\, \{H_{k+1}(X; \; \mathscr{F}) \to H_{k+1}(X_j; \; \mathscr{F})\}.$$

This is necessary and sufficient for the map (5) to be an isomorphism of topological vector spaces.

Proof. Under the assumptions made the spaces $H_c{}^{k+1}(X_j; \; \mathscr{F})$ $(j = 1, 2, \ldots)$ are separated (cf. the corollary to Theorem 3.3 of Chapter 3). Moreover, it is obvious that the polar of the set $\mathrm{Im}\, \{H_{k+1}(X_{j'}; \; \mathscr{F}) \to H_{k+1}(X_j; \; \mathscr{F})\}$ in $H_c{}^{k+1}(X_j; \; \mathscr{F})$ coincides with the kernel $\mathrm{Ker}\, \{H_c{}^{k+1}(X_j; \; \mathscr{F}) \to H_c{}^{k+1}(X_{j'}; \; \mathscr{F})\}$. According to Theorem 4.4, the condition of Theorem 4.4.3 is equivalent to the separatedness of the space $H_c{}^{k+1}(X; \; \mathscr{F})$. The theorem is proved.

4.4.4. Corollary. Let the topological vector spaces $H_k(X_j; \; \mathscr{F})$ $(j = 1, 2, \ldots)$ be separated. Then the space $H_k(X; \; \mathscr{F})$ is separated [and the map (5) is an isomorphism], if the following condition holds: for each $j = 1, 2, \ldots$ the image of the map

$$H_{k+1}(X; \; \mathscr{F}) \to H_{k+1}(X_j; \; \mathscr{F})$$

is everywhere dense in the image of the map

$$H_{k+1}(X_{j+1}; \; \mathscr{F}) \to H_{k+1}(X_j; \; \mathscr{F}).$$

4.4.5.   Theorem. Let the spaces $H_k(X_j; \mathscr{F})$ $(j = 1, 2, \ldots)$ be separated and suppose for each $j = 1, 2, \ldots$ the image of the space $H_{k+1}(X_{j+1}; \mathscr{F})$ is everywhere dense in the space $H_{k+1}(X_j; \mathscr{F})$. Then the space $H_k(X; \mathscr{F})$ is separated and, consequently, the map (5) is an isomorphism of topological vector spaces.[13]

Proof. Under the assumptions made, the spaces $H_c^{k+1}(X_j; \mathscr{F})$ $(j = 1, 2, \ldots)$ are separated and the maps $H_c^{k+1}(X_j; \mathscr{F}) \to H_c^{k+1}(X_{j+1}; \mathscr{F})$ are injective. By Corollary 4.4.2 the space $H_c^{k+1}(X; \mathscr{F})$ is also separated (cf. the corollary to Theorem 3.3 of Chapter 3). The theorem is proved.

5.   Tests for Separatedness

5.1. Theorem. Let X be a complex space which is countable at infinity, and $\mathscr{F}$ be a coherent analytic sheaf on X. Then the topological vector space $H_k^c(X; \mathscr{F})$ is separated if and only if the following condition holds: for each relatively compact open set U in X and each bounded set B in $H_k^c(U; \mathscr{F})$, whose image in the space $H_k^c(X; \mathscr{F})$ is equal to zero, there exists a relatively compact open set U' in X, containing U, and such that the image of the set B in the space $H_k^c(U'; \mathscr{F})$ is equal to zero.[14]

Proof. Let $\mathfrak{U} = (U_i)_{i \geq 1}$ be a locally finite covering of the space X by relatively compact holomorphically complete open sets. Let $\mathfrak{U}_p = \{U_1, \ldots, U_p\}$ for each $p = 1, 2, \ldots$ Since $\mathfrak{U} = \cup \, \mathfrak{U}_p$, the space of chains $C_k^c(\mathfrak{U}; \mathscr{H}_0^c(\mathscr{F}))$ can be identified naturally with the strict locally convex inductive limit of the spaces $C_k^c(\mathfrak{U}_p; \mathscr{H}_0^c(\mathscr{F}))$ $(p = 1, 2, \ldots)$ and is the strong dual to a Frechet–Schwartz space. Let B be a closed convex balanced bounded set in $C_k^c(\mathfrak{U}; \mathscr{H}_0^c(\mathscr{F}))$. We denote by E_B the vector space generated in $C_k^c(\mathfrak{U}; \mathscr{H}_0^c(\mathscr{F}))$ by the set B and endowed with the norm defined by this set (by virtue of Kolmogorov's theorem). Then the space E_B is complete in the norm topology. The space $C_k^c(\mathfrak{U}; \mathscr{H}_0^c(\mathscr{F}))$ can be represented as the locally convex inductive limit of a sequence of Banach spaces E_B, which is increasing by inclusion, with completely continuous maps:

$$C_k^c(\mathfrak{U}; \mathscr{H}_0^c(\mathscr{F})) = \varinjlim E_B.$$

According to a theorem of Sebastian-i-Silva the set M in $C_k^c(\mathfrak{U}; \mathscr{H}_0^c(\mathscr{F}))$ is closed if and only if for each closed convex balanced bounded set B in $C_k^c(\mathfrak{U}; \mathscr{H}_0^c(\mathscr{F}))$ the intersection $M \cap E_B$ is closed in the space E_B.[15]

5.1.1. Let us assume that the condition of the theorem holds. We show that then the space $H_k{}^c(X;\ \mathscr{F})$ is separated, i.e., that the subspace $\partial C_{k+1}{}^c(\mathfrak{U};\ \mathscr{H}_0{}^c(\mathscr{F}))$ in $C_k{}^c(\mathfrak{U};\ \mathscr{H}_0{}^c(\mathscr{F}))$ is closed (cf. point 2.2.2 of Chapter 3). Let B be an arbitrary closed convex balanced bounded set in $C_k{}^c(\mathfrak{U};\ \mathscr{H}_0{}^c(\mathscr{F}))$. It suffices to prove that the intersection $\partial C_{k+1}{}^c(\mathfrak{U};\ \mathscr{H}_0{}^c(\mathscr{F})) \cap E_B$ is closed in E_B. On the other hand, for some positive integer p_0 the set B is contained in $C_k{}^c(\mathfrak{U}_{p_0};\ \mathscr{H}_0{}^c(\mathscr{F}))$. Since the image of the intersection $B \cap \partial C_{k+1}{}^c(\mathfrak{U};\ \mathscr{H}_0{}^c(\mathscr{F}))$ in the space $H_k{}^c(\mathfrak{U};\ \mathscr{H}_0{}^c(\mathscr{F}))$ is equal to zero, by the condition of the theorem there exists a $p > p_0$ such that the image of the intersection $B \cap \partial C_{k+1}{}^c(\mathfrak{U};\ \mathscr{H}_0{}^c(\mathscr{F}))$ in the space $H_k{}^c(\mathfrak{U}_p;\ \mathscr{H}_0{}^c(\mathscr{F}))$ is equal to zero. The latter means that the intersection $B \cap \partial C_{k+1}{}^c(\mathfrak{U};\ \mathscr{H}_0{}^c(\mathscr{F}))$ is contained in $\partial C_{k+1}{}^c(\mathfrak{U}_p;\ \mathscr{H}_0{}^c(\mathscr{F}))$, so

$$E_B \cap \partial C_{k+1}^c(\mathfrak{U};\ \mathscr{H}_0^c(\mathscr{F})) = E_B \cap \partial C_{k+1}^c(\mathfrak{U}_p;\ \mathscr{H}_0^c(\mathscr{F})).$$

Let $R_1 = \operatorname{Ker} r_1$ be the kernel of the continuous linear map

$$r_1\colon C_{k+1}^c(\mathfrak{U}_p;\ \mathscr{H}_0^c(\mathscr{F}))/Z_1 \times E_B \to C_k^c(\mathfrak{U};\ \mathscr{H}_0^c(\mathscr{F})),$$

where $Z_1 = Z_{k+1}{}^c(\mathfrak{U}_p;\ \mathscr{H}_0{}^c(\mathscr{F}))$ is the kernel of the boundary operator ∂, and $C_{k+1}{}^c(\mathfrak{U}_p;\ \mathscr{H}_0{}^c(\mathscr{F})) \to C_k{}^c(\mathfrak{U}_p;\ \mathscr{H}_0{}^c(\mathscr{F}))$ and $r_1(f, g) = \partial f - g$ for any $f \in C_{k+1}{}^c(\mathfrak{U}_p;\ \mathscr{H}_0{}^c(\mathscr{F}))$, $g \in E_B$. Let $\mathfrak{V} = (V_i)_{i \geq 1}$ be a locally finite covering of the space X by relatively compact holomorphically complete open sets, for which $\overline{U_i} \subset V_i$ ($i = 1, 2, \ldots$). Analogously to the preceding, we define the kernel R_2 of the continuous linear map

$$r_2\colon C_{k+1}^c(\mathfrak{V}_p;\ \mathscr{H}_0^c(\mathscr{F}))/Z_2 \times E_B \to C_k^c(\mathfrak{V};\ \mathscr{H}_0^c(\mathscr{F})),$$

where $Z_2 = Z_{k+1}{}^c(\mathfrak{V}_p;\ \mathscr{H}_0{}^c(\mathscr{F}))$ is the kernel of the boundary operator ∂: $C_{k+1}{}^c(\mathfrak{V}_p;\ \mathscr{H}_0{}^c(\mathscr{F})) \to C_k{}^c(\mathfrak{V}_p;\ \mathscr{H}_0{}^c(\mathscr{F}))$, and $r_2(f, g) = \partial f - g$ for any $f \in C_{k+1}{}^c(\mathfrak{V}_p;\ \mathscr{H}_0{}^c(\mathscr{F}))$, $g \in E_B$. The natural injective maps

$$H_0^c(U_{i_0 \ldots i_{k+1}};\ \mathscr{F}) \to H_0^c(V_{i_0 \ldots i_{k+1}};\ \mathscr{F})$$

are completely continuous and define an injective completely continuous linear map

$$\theta_1\colon C_{k+1}^c(\mathfrak{U}_p;\ \mathscr{H}_0^c(\mathscr{F}))/Z_1 \to C_{k+1}^c(\mathfrak{V}_p;\ \mathscr{H}_0^c(\mathscr{F}))/Z_2.$$

Let $\theta_2\colon E_B \to E_B$ be the identity isomorphism. Then the continuous linear map

$$\theta = (\theta_1, \theta_2): \ R_1 \to R_2$$

is bijective and, consequently, is an isomorphism of topological vector spaces (by the open mapping theorem). In fact, the surjectivity of the map θ follows from the condition of the theorem and the injectivity is obvious. Since

$$\theta = (\theta_1, \ 0) + (0, \ \theta_2)$$

and θ_1 is completely continuous, by a theorem of Schwartz [1] the map θ_2 has closed image

$$\theta_2(R_1) = E_B \cap \partial C^c_{k+1}(\mathfrak{U}_p; \ \mathscr{H}^c_0(\mathscr{F}))$$

in E_B. Thus the sufficiency of the condition of the theorem is proved.

5.1.2. We show that conversely, too, the condition of the theorem holds if the space $H_k{}^c(X; \ \mathscr{F})$ is separated. Let $\mathfrak{U} = (U_i)_{i \geq 1}$ be a locally finite covering of the space X by relatively compact holomorphically complete open sets. Since the space $H_k{}^c(\mathfrak{U}; \ \mathscr{H}_0{}^c(\mathscr{F}))$ is separated, the subspace $\partial C_{k+1}{}^c(\mathfrak{U}; \ \mathscr{H}_0{}^c(\mathscr{F}))$ is closed in the space $C_k{}^c(\mathfrak{U}; \ \mathscr{H}_0{}^c(\mathscr{F}))$ and, consequently, is the strong dual to a Frechet–Schwartz space. For each $p = 1, 2, \ldots$ we endow the vector space $\partial C_{k+1}{}^c(\mathfrak{U}_p; \ \mathscr{H}_0{}^c(\mathscr{F}))$ with the strongest locally convex topology for which the boundary operator

$$\partial: \ C^c_{k+1}(\mathfrak{U}_p; \ \mathscr{H}^c_0(\mathscr{F})) \to \partial C^c_{k+1}(\mathfrak{U}_p; \ \mathscr{H}^c_0(\mathscr{F}))$$

is continuous. Then the natural map

$$C^c_{k+1}(\mathfrak{U}_p; \ \mathscr{H}^c_0(\mathscr{F}))/Z^c_{k+1}(\mathfrak{U}_p; \ \mathscr{H}^c_0(\mathscr{F})) \to \partial C^c_{k+1}(\mathfrak{U}_p; \ \mathscr{H}^c_0(\mathscr{F}))$$

is an isomorphism of topological vector spaces (by definition of the topology on the quotient space). Consequently, the space $\partial C_{k+1}{}^c(\mathfrak{U}_p; \ \mathscr{H}_0{}^c(\mathscr{F}))$ is the strong dual to a Frechet–Schwartz space. Hence the space $\partial C_{k+1}{}^c(\mathfrak{U}; \ \mathscr{H}_0{}^c(\mathscr{F}))$ can be identified naturally with the locally convex limit of the spaces $\partial C_{k+1}{}^c(\mathfrak{U}_p; \ \mathscr{H}_0{}^c(\mathscr{F}))$ $(p = 1, 2, \ldots)$. Let B_0 be a bounded set in the space $H_k{}^c(\mathfrak{U}_p; \ \mathscr{H}_0{}^c(\mathscr{F}))$, whose image in the space $H_k{}^c(\mathfrak{U}; \ \mathscr{H}_0{}^c(\mathscr{F}))$ is equal to zero. Since the space $Z_k{}^c(\mathfrak{U}_p; \ \mathscr{H}_0{}^c(\mathscr{F}))$ is the strong dual to a Frechet–Schwartz space, there exists in it a bounded set B, whose image in the space $H_k{}^c(\mathfrak{U}_p; \ \mathscr{H}_0{}^c(\mathscr{F}))$ is equal to B_0. Since the image of the set B in the space $H_k{}^c(\mathfrak{U}; \ \mathscr{H}_0{}^c(\mathscr{F}))$ is equal to zero, $B \subset \partial C_{k+1}{}^c(\mathfrak{U}; \ \mathscr{H}_0{}^c(\mathscr{F}))$.

Consequently, there exists a $p' \geq p$, such that the set B is contained in $\partial C_{k+1}{}^c(\mathfrak{U}_{p'}; \mathscr{H}_0{}^c(\mathscr{F}))$. This means that the image of the set B_0 in the space $H_k{}^c(\mathfrak{U}_{p'}; \mathscr{H}_0{}^c(\mathscr{F}))$ is equal to zero. The theorem is completely proved.

5.1.3. Corollary. The space $H_k{}^c(X; \mathscr{F})$ is separated if the following condition holds: for each relatively compact open set U in X there exists a relatively compact open set U' in X, containing U and such that the kernel of the map

$$H_k^c(U; \mathscr{F}) \to H_k^c(X; \mathscr{F})$$

is contained in the kernel of the map[16,17]

$$H_k^c(U; \mathscr{F}) \to H_k^c(U'; \mathscr{F}).$$

5.2. Theorem. Let X be a complex space which is countable at infinity, and $\mathscr{F}$ be a coherent analytic sheaf on X. Then the topological vector space $H_c{}^k(X; \mathscr{F})$ is separated if and only if the following condition holds: for each relatively compact open set U in X and each bounded set B in $H_c{}^k(U; \mathscr{F})$ whose image in the space $H_c{}^k(X; \mathscr{F})$ is equal to zero, there exists a relatively compact open set U' in X, containing U and such that the image of the set B in the space $H_c{}^k(U'; \mathscr{F})$ is equal to zero.[18]

Proof. We give a sequence of coverings $\mathfrak{M}_i$ ($i = 1, 2, \ldots$) of the space X, satisfying the conditions of points 3.6 and 3.7. We define the cochain spaces $C_X{}^k(\mathscr{F})$ and $C_U{}^k(\mathscr{F})$ for an arbitrary open set $U \subset X$ (cf. point 3.6). They are strong duals to Frechet–Schwartz spaces. Here the topology of the space $C_X{}^k(\mathscr{F})$ coincides with the topology of the locally convex inductive limit of the spaces $C_U{}^k(\mathscr{F})$ with respect to the filtered set, ordered by inclusion, of all relatively compact open sets U in X. Moreover, for any open set $U \subset X$ there is a natural isomorphism of topological vector spaces

$$H_c^k(U; \mathscr{F}) = H^k C_U^{\cdot}(\mathscr{F})$$

(cf. point 3.6.4). Let B be a closed convex balanced bounded set in $C_X{}^k(\mathscr{F})$. We denote by E_B the vector subspace of $C_X{}^k(\mathscr{F})$, generated by the set B and endowed with the norm, defined by this set. The space E_B is complete in the norm topology. Since the space $C_X{}^k(\mathscr{F})$ is the strong dual to a Frechet–Schwartz space, it can be represented in the form of a locally convex inductive limit of an increasing sequence of Banach spaces E_B with completely continuous maps:

$$C_X^k(\mathcal{F}) = \varinjlim E_B.$$

By the theorem of Sebastian-i-Silva the set M in the space $C_X^k(\mathcal{F})$ is closed if and only if, for each closed convex balanced bounded set B in $C_X^k(\mathcal{F})$ the intersection $M \cap E_B$ is closed in the space E_B.

5.2.1. Let us assume that the condition of the theorem holds. We show that then the space $H_c^k(X: \mathcal{F})$ is separated, i.e., that the subspace $\delta C_X^{k-1}(\mathcal{F})$ in $C_X^k(\mathcal{F})$ is closed. Let B be an arbitrary closed convex balanced bounded set in $C_X^k(\mathcal{F})$. It suffices to prove that the intersection $E_B \cap \delta C_X^{k-1}(\mathcal{F})$ is closed in E_B. On the other hand, for some integer $i > 0$ and some relatively compact open U in X the set B is contained in $C_U^k(\mathfrak{M}_i; \mathcal{F})$. Since the image of the intersection $B \cap \delta C_X^{k-1}(\mathcal{F})$ in the space $H_c^k(X: \mathcal{F})$ is equal to zero, by the condition of the theorem there exists a relatively compact open set U' in X, containing U, such that the image of the intersection $B \cap \delta C_X^{k-1}(\mathcal{F})$ in the space $H_c^k(U': \mathcal{F})$ is equal to zero. We consider the compact set $S = U'$ in X. It is obvious that under the composite map $H_c^k(U: \mathcal{F}) \to H_c^k(U': \mathcal{F}) \to H_S^k(X; \mathcal{F})$ the image of the intersection $B \cap \delta C_X^{k-1}(\mathcal{F})$ in the space $H_S^k(X; \mathcal{F})$ is equal to zero. Let $\mathfrak{U} = (U_j)$ be a finite covering of the set S by open sets. We shall assume that each set $M_{j_0 \ldots j_k}$, where $M_{j_0}, \ldots, M_{j_k} \in \mathfrak{M}_i$, coincides with the closure of its interior (i.e., is a canonical set; cf. Aleksandrov [2]); we set $U_k = M_j$. Thus, $\mathfrak{U}$ consists of relatively compact holomorphically complete open sets. Let $\mathfrak{B} = (V_j)$ be a finite covering of the set $\cup U_j \setminus S$ by relatively compact sets in X, which are holomorphically complete, inscribed in $\mathfrak{U}$. We define a complex

$$K^k(\mathcal{F}) = C^k(\mathfrak{U}; \mathcal{F}) \times C^{k-1}(\mathfrak{B}; \mathcal{F})$$

(cf. point 3.1). Then there is defined a natural map of cochain complexes

$$\alpha \colon C_U^k(\mathfrak{M}_i; \mathcal{F}) \to K^k(\mathcal{F}),$$

for which $\alpha(f) = (f_1, f_2)$, where $f \in C_U^k(\mathfrak{M}_i; \mathcal{F})$, $f_1 \in C^k(\mathfrak{U}; \mathcal{F})$ is the restriction of f to $\mathfrak{U}$, $f_2 = 0$. Since $U \cap V_j = \varnothing$, α is compatible with the coboundary operators. It is also obvious that α is injective. Since the image of the intersection $B \cap \delta C_X^{k-1}(\mathcal{F})$ is equal to zero in $H_S^k(X; \mathcal{F})$, one has $\alpha(B \cap \delta C_X^{k-1}(\mathcal{F})) \subset \delta K^{k-1}(\mathcal{F})$. Since the image of the set $\delta K^{k-1}(\mathcal{F})$ in the space $H_c^k(X; \mathcal{F})$ is equal to zero, we get

$$\alpha (E_B \cap \delta C_X^{k-1}(\mathcal{F})) = \alpha (E_B) \cap \delta K^{k-1}(\mathcal{F}).$$

Let $R = \operatorname{Ker} r$ be the kernel of the continuous linear map

$$r: K^{k-1}(\mathcal{F})/Z \times E_B \rightarrow K^k(\mathcal{F}),$$

where Z is the kernel of the coboundary operator δ: $K^{k-1}(\mathcal{F}) \rightarrow K^k(\mathcal{F})$, and $r(f, g) = \delta f - \alpha g$ for any $f \in K^{k-1}(\mathcal{F})$, $g \in E_B$. Let $\mathfrak{U}_1 = (U_{1j})$ and $\mathfrak{V}_1 = (V_{1j})$ be finite coverings, respectively, of the set S and $\cup U_{1j} \setminus S$ by relatively compact holomorphically complete open sets, where $\mathfrak{V}_1$ is inscribed in $\mathfrak{U}_1$. We shall assume that $\bar{U}_{1j} \subset U_j$, $\bar{V}_{1j} \subset V_j$, and the restriction maps

$$\Gamma (U_{i_0 \dots i_{k-1}}; \mathcal{F}) \rightarrow \Gamma (U_{1i_0 \dots i_{k-1}}; \mathcal{F}),$$
$$\Gamma (V_{i_0 \dots i_{k-2}}; \mathcal{F}) \rightarrow \Gamma (V_{1i_0 \dots i_{k-2}}; \mathcal{F})$$

are injective. Analogously to the preceding, we define the kernel $R_1 = \operatorname{Ker} r_1$ of the continuous linear map

$$r_1: K_1^{k-1}(\mathcal{F})/Z_1 \times E_B \rightarrow K_1^k(\mathcal{F}),$$

where $K_1{}^k(\mathcal{F}) = C^k(\mathfrak{U}_1; \mathcal{F}) \times C^{k-1}(\mathfrak{V}_1; \mathcal{F})$, Z_1 is the kernel of the coboundary operator δ: $K_1{}^{k-1}(\mathcal{F}) \rightarrow K_1{}^k(\mathcal{F})$, and $r_1(f, g) = \delta f - \alpha g$ for any $f \in K^{k-1}(\mathcal{F})$, $g \in E_B$. The restriction maps define an injective completely continuous map

$$\theta_1: K^{k-1}(\mathcal{F})/Z \rightarrow K_1^{k-1}(\mathcal{F})/Z_1.$$

Let θ_2: $E_B \rightarrow E_B$ be the identity isomorphism. Then the continuous linear map

$$\theta = (\theta_1, \theta_2): R \rightarrow R_1$$

is bijective and consequently is an isomorphism of topological vector spaces (by the open map theorem). In fact, the surjectivity of the map θ follows from the condition of the theorem, while the injectivity is obvious. Since $\theta = (\theta_1, 0) + (0, \theta_2)$, and θ_1 is completely continuous, by a theorem of Schwartz [1] the map θ_2 has closed image in E_B:

$$\theta_2 (R) = \alpha (E_B) \cap \delta K^{k-1}(\mathcal{F}).$$

This means that the intersection $E_B \subset \delta C_X^{k-1}(\mathscr{F})$ is closed in the space E_B. The sufficiency of the condition is proved.

5.2.2. We show that conversely, too, the condition of the theorem holds, if the space $H_c^k(X; \mathscr{F})$ is separated. It follows from the separatedness of the space $H_c^k(X; \mathscr{F})$ that the subspace $\delta C_X^{k-1}(\mathscr{F})$ is closed in the space $C_X^k(\mathscr{F})$ and, consequently, is the strong dual to a Frechet–Schwartz space. For each relatively compact open set U in X we endow the vector space $\delta C_U^{k-1}(\mathscr{F})$ with the strongest locally convex topology for which the coboundary operator $\delta \colon C_U^{k-1}(\mathscr{F}) \to \delta C_U^{k-1}(\mathscr{F})$ is continuous. Then the natural map $C_U^{k-1}(\mathscr{F})/Z_U^{k-1}(\mathscr{F}) \to \delta C_U^{k-1}(\mathscr{F})$, defined by the operator δ, is an isomorphism of topological vector spaces. Consequently, in this topology $\delta C_U^{k-1}(\mathscr{F})$ is the strong dual to a Frechet–Schwartz space. Hence the space $\delta C_X^{k-1}(\mathscr{F})$ can be identified naturally with the locally convex inductive limit of the spaces $\delta C_U^{k-1}(\mathscr{F})$ with respect to the filtered set, ordered by inclusion, of relatively compact open subsets U in X. Let B_0 be a bounded set in the space $H_c^k(U; \mathscr{F})$, whose image in the space $H_c^k(X; \mathscr{F})$ is equal to zero. Then there exists a bounded set B in the space

$Z_U^k(\mathscr{F}) = \mathrm{Ker}\, \{C_U^k(\mathscr{F}) \xrightarrow{\delta} C_U^{k+1}(\mathscr{F})\}$ whose image in the space $H_c^k(U; \mathscr{F}) = H^k C_U^{\cdot}(\mathscr{F})$ is equal to B_0. Since the image of the set B in the space $H_c^k(X; \mathscr{F})$ is equal to zero, $B \subset \delta C_X^{k-1}(\mathscr{F})$. Consequently, there exists a relatively compact open set U' in X, containing U, such that the set B is contained in $\delta C_{U'}^{k-1}(\mathscr{F})$. This means that the image of the set B_0 in the space $H_c^k(U'; \mathscr{F})$ is equal to zero. The theorem is proved.[19]

5.2.3. Corollary. The space $H_c^k(X; \mathscr{F})$ is separated if the following condition holds: for each relatively compact open set U in X there exists a relatively compact open set U' in X, containing U and such that the kernel of the map $H_c^k(U; \mathscr{F}) \to H_c^k(X; \mathscr{F})$ is contained in the kernel of the map $H_c^k(U; \mathscr{F}) \to H_c^k(U'; \mathscr{F})$.[20]

Chapter 5

APPLICATIONS

1. Dimension of Support

1.1. Let S be an arbitrary set in the complex space X. By the *dimension* $\dim_x S$ of the set S at the point $x \in X$ is meant the smallest integer d such that there exists a germ of an analytic set at the point x of dimension d, containing the germ S_x of the set S. The number

$$\dim S = \sup_{x \in X} \dim_x S$$

is called *the dimension of the set S*.[1]

Theorem. Let S be an arbitrary closed set of dimension d in the complex space X. Then for any analytic sheaf $\mathscr{F}$ on X

$$H_k^S (X; \mathscr{F}) = 0 \quad \text{for} \quad k > d.$$

Proof. We make use of the spectral sequence

$$E_{p,\,q}^2 = H^{-p} (X; \mathscr{H}_q^S (\mathscr{F})) \Longrightarrow H_{p+q}^S (X; \mathscr{F})$$

(cf. Theorem 5.3). It suffices to prove that

$$E_{p,\,q}^2 = 0 \quad \text{for} \quad p+q > d.$$

This is obvious for $p > 0$. If $p \leq 0$, then $q > d$. Consequently, it suffices to prove that

$$\mathcal{H}_q^S(\mathcal{F}) = 0 \quad \text{for} \quad q > d.$$

Since the last assertion has local character, one can assume that S is an analytic set in a domain G of the space $\mathbf{C}^n$. Let S_0 be the set of singular points of the analytic set S. Then there exists an exact sequence

$$\ldots \to H_k^{S_0}(G; \mathcal{F}^G) \to H_k^S(G; \mathcal{F}^G) \to H_k^{S \setminus S_0}(G \setminus S_0; \mathcal{F}^G) \to \ldots$$

(cf. point 2.6 of Chapter 2). Let us assume that

$$H_k^{S \setminus S_0}(G \setminus S_0; \mathcal{F}^G) = 0 \quad \text{for} \quad k > d.$$

Then, applying induction on d, we get

$$H_k^S(G; \mathcal{F}^G) = 0 \quad \text{for} \quad k > d.$$

Thus, it suffices to prove that

$$\mathcal{H}_k^S(\mathcal{F}) = 0 \quad \text{for} \quad k > d,$$

where S is an analytic set in the domain G, defined by the equations

$$z_{d+1} = z_{d+2} = \ldots = z_n = 0.$$

By definition of the homology sheaf of germs

$$\mathcal{H}_k^S(\mathcal{F}) = \mathcal{E}\mathrm{xt}_{\mathcal{O}_G}^{n-k}(\mathcal{F}_S^G, \Omega_G^n)$$

(cf. point 1.3 of Chapter 2). On the other hand, the natural projection $p \colon G \to S$ induces the identity isomorphism of the analytic set S onto itself. The sheaf $\mathcal{F}_S$ is endowed with a natural $\mathcal{O}_S$-module structure. Consequently, p induces a natural isomorphism of analytic sheaves

$$\mathcal{E}\mathrm{xt}_{\mathcal{O}_G}^{n-k}(\mathcal{F}_S^G, \Omega_G^n) = \mathcal{E}\mathrm{xt}_{\mathcal{O}_S}^{d-k}(\mathcal{F}_S, \Omega_S^d)$$

(cf. Theorem 1.2 of Chapter 2). The theorem is completely proved.[2]

1.1.1. Corollary. The natural map

$$H_k(X; \mathcal{F}) \to H_k(X \setminus S; \mathcal{F})$$

is injective for $k = d + 1$ and bijective for $k > d + 1$.

1.1.2. Corollary. Let Φ be an arbitrary family of supports in X. Then if $\dim S \leq d$ for each $S \in \Phi$, then

$$H_k^\Phi(X; \mathscr{F}) = 0 \quad \text{for} \quad k > d.$$

1.1.3. By the *support* of the analytic sheaf $\mathscr{F}$ is meant the smallest closed set Supp $\mathscr{F}$, containing all points $x \in X$, for which $\mathscr{F}_x \neq 0$.[3]

Corollary. For any family of supports Φ in X

$$H_k^\Phi(X; \mathscr{F}) = 0 \quad \text{for} \quad k > d,$$

where d is the dimension of the support Supp $\mathscr{F}$.[4]

1.2. Theorem. Let the complex space X of dimension n, which is countable at infinity, contain only a finite number of compact irreducible components. Then for any coherent analytic sheaf $\mathscr{F}$ on X the topological vector spaces $H_{n-1}{}^c(X; \mathscr{F})$ and $H^n(X; \mathscr{F})$ are separated.

Proof. We make use of the spectral sequence

$$E_{p,q}^2 = H_c^{-p}(X; \mathscr{H}_q(\mathscr{F})) \Rightarrow H_{p+q}^c(X; \mathscr{F})$$

(cf. Theorem 4.1 of Chapter 2). By Theorem 1.1

$$\mathscr{H}_q(\mathscr{F}) = 0 \quad \text{for} \quad q > n.$$

Consequently, $E_{p,n-p}^2 = 0$ for $p \neq 0$, i.e., there exists a natural isomorphism

$$H_n^c(X; \mathscr{F}) = \Gamma_c(X; \mathscr{H}_n(\mathscr{F})).$$

If a neighborhood of an arbitrary point of the space X is realized as an analytic set in a domain G of the space $\mathbf{C}^m$, then

$$\mathscr{H}_n(\mathscr{F}) = \mathscr{E}\mathrm{xt}_{\mathcal{O}_G}^{m-n}(\mathscr{F}^G, \Omega_G^m)$$

(cf. point 1.3 of Chapter 2). In particular, in a neighborhood of a regular point

$$\mathscr{H}_n(\mathscr{F}) = \mathscr{H}\mathrm{om}_{\mathcal{O}_G}(\mathscr{F}^G, \Omega_G^n).$$

Consequently, the coherent analytic sheaf $\mathscr{H}_n(\mathscr{F})$ has no torsion. Since the space X contains only a finite number of compact irreducible components, the vector space $\Gamma_c(X; \mathscr{H}_n(\mathscr{F}))$ is finite-dimensional. Let U be an arbitrary relatively compact open set in X and let U' be a relatively compact open set in X, containing U and all compact irreducible components of the space X. Then

$$\Gamma_c(X \backslash U'; \mathscr{H}_n(\mathscr{F})) = 0.$$

Hence, from the exact sequence

$$\ldots \rightarrow H_n^c(X \backslash U'; \mathscr{F}) \rightarrow H_{n-1}^c(U'; \mathscr{F}) \rightarrow H_{n-1}^c(X; \mathscr{F}) \rightarrow \ldots$$

we get that the kernel of the map

$$H_{n-1}^c(U; \mathscr{F}) \rightarrow H_{n-1}^c(X; \mathscr{F})$$

coincides with the kernel of the map

$$H_{n-1}^c(U; \mathscr{F}) \rightarrow H_{n-1}^c(U'; \mathscr{F}).$$

By Corollary 5.1.3 of Chapter 4, the space $H_{n-1}{}^c(X; \mathscr{F})$ is separated. Then the space $H^n(X; \mathscr{F})$ is also separated by the corollary to Theorem 2.3 of Chapter 3. The theorem is proved.

1.2.1. Corollary. If the space X contains only a finite number of compact irreducible components, then the space $H^n(X; \mathscr{F})$ is finite-dimensional. If the space X does not contain any compact irreducible components, then $H^n(X; \mathscr{F}) = 0.$[5]

Proof. Under the assumptions made, the vector space

$$H_n^c(X; \mathscr{F}) = \Gamma_c(X; \mathscr{H}_n(\mathscr{F}))$$

is finite-dimensional or respectively trivial. Hence the assertion follows from Theorem 2.3 of Chapter 3.

1.3. Theorem. Let the coherent analytic sheaf $\mathscr{F}$ on the complex space X, which is countable at infinity, have only a finite number of linearly independent sections with compact supports. Then the topological vector spaces $H_c^1(X; \mathscr{F})$ and $H_0(X; \mathscr{F})$ are separated.

Proof. Let U be an arbitrary relatively compact open set in X and let U' be a relatively compact open set in X, containing U and those supports of sections of the sheaf $\mathscr{F}$, which are compact. Then

$$\Gamma_c(X \backslash U'; \mathscr{F}) = 0.$$

Consequently, from the exact sequence

$$\ldots \to \Gamma_c(X \setminus U'; \mathscr{F}) \to H_c^1(U'; \mathscr{F}) \to H_c^1(X; \mathscr{F}) \to \ldots$$

we get that the kernel of the map

$$H_c^1(U; \mathscr{F}) \to H_c^1(X; \mathscr{F})$$

coincides with the kernel of the map

$$H_c^1(U; \mathscr{F}) \to H_c^1(U'; \mathscr{F}).$$

Thus by Corollary 5.2.3 of Chapter 4, the space $H_c{}^1(X; \mathscr{F})$ is separated. From this, in view of the corollary to Theorem 3.3 of Chapter 3, the space $H_0(X; \mathscr{F})$ is also separated. The theorem is proved.

1.3.1. Corollary. If the sheaf $\mathscr{F}$ has only a finite number of linearly independent sections with compact supports, then the vector space $H_0(X; \mathscr{F})$ is finite-dimensional. If the sheaf $\mathscr{F}$ has no sections with compact supports, then $H_0(X; \mathscr{F}) = 0$.

Proof. Under the assumptions made, the vector space $\Gamma_c(X; \mathscr{F})$ is finite-dimensional or respectively trivial. Hence the assertion follows from Theorem 3.3 of Chapter 3.

2. Homological Codimension

2.1. Let X be a complex space, $\mathscr{F}$ be a coherent analytic sheaf on X, and x be an arbitrary point in X, a neighborhood of which is realized as an analytic set in a domain G of the space $\mathbf{C}^n$. By the *homological codimension* of the sheaf $\mathscr{F}$ at the point x we mean the number

$$\operatorname{codh}_x \mathscr{F} = n - h,$$

where h is the homological (i.e, projective) dimension of the stalk $\mathscr{F}_x$ over the ring $\mathcal{O}_{G, x}$. In fact, this definition is independent of the choice of realization of a neighborhood of the point x as an analytic set in a domain of a complex number space (cf. Andreotti [1] and Grauert [3]). By the *homological codimension* of the sheaf $\mathscr{F}$ on X is meant the number[6,7]

$$\operatorname{codh} \mathscr{F} = \min_{x \in X} \operatorname{codh}_x \mathscr{F} ,$$

Theorem. Let X be a complex space which is countable at infinity, $\mathscr{F}$ be a coherent analytic sheaf on X of homological codimension codh $\mathscr{F} = h$, and S be a compact set in X, for which there exists a fundamental system of holomorphically complete open neighborhoods. Then[8]

$$H_S^k(X; \mathscr{F}) = 0 \quad \text{for} \quad k < h.$$

Proof. We use the spectral sequence

$$E_{p,q}^2 = H^{-p}(S; \mathscr{H}_q(\mathscr{F})) \Rightarrow H_{p+q}(S; \mathscr{F})$$

(cf. Theorem 4.1 of Chapter 2). Since $E_{p,q}2 = 0$ for $p \neq 0$, we get a natural isomorphism of vector spaces

$$H_k(S; \mathscr{F}) = \Gamma(S; \mathscr{H}_k(\mathscr{F})).$$

If a neighborhood of an arbitrary point of the space X can be realized as an analytic set in a domain G of the space $\mathbf{C}^n$, then in this neighborhood

$$\mathscr{H}_k(\mathscr{F}) = \mathscr{E}\mathrm{xt}_{\mathcal{O}_G}^{n-k}(\mathscr{F}^G, \Omega_G^n)$$

(cf. point 1.3 of Chapter 2); consequently, $\mathscr{H}_k(\mathscr{F}) = 0$ for $k < h$. The assertion follows from this by virtue of Theorem 3.5 of Chapter 4. The theorem is proved.

Corollary. The natural map

$$H^k(X; \mathscr{F}) \to H^k(X \backslash S; \mathscr{F})$$

is bijective for $k < h - 1$ and injective for $k = h - 1$.[9]

2.2. Theorem. Let X be a complex space which is countable at infinity, $\mathscr{F}$ be a coherent analytic sheaf on X of homological codimension codh $\mathscr{F} = h$, and S be an analytic set in X of dimension d. Then

$$H_S^k(X; \mathscr{F}) = 0 \quad \text{for} \quad k < h - d.$$

Proof. We make use of the spectral sequence

$$E_{p,q}^2 = H_c^{-p}(S; \mathscr{H}_q(\mathscr{F})) \Rightarrow H_{p+q}^c(S; \mathscr{F})$$

(cf. Theorem 4.1 of Chapter 2). Since $\mathscr{H}_q(\mathscr{F}) = 0$ for $q < h$, one has $E_{p,q}2 = 0$ for $q < h$. By Reiffen's theorem [1], $E_{p,q}2 = 0$ for $p < -d$ (cf. Note 2). Consequently,

$$H_k^c(S; \mathcal{F}) = 0 \quad \text{for} \quad k < h - d.$$

Now from this it follows that the topological vector spaces $H_S{}^k(X; \mathcal{F})$ are separated for $k \leq h - d$ (cf. the corollary to Theorem 3.3 of Chapter 4). The assertion follows now from Theorem 3.3 of Chapter 4. The theorem is proved.

Corollary. The natural map

$$H^k(X; \mathcal{F}) \to H^k(X \backslash S; \mathcal{F})$$

is bijective for $k < h - d - 1$ and injective for $k = h - d - 1$.[10]

3. The Results of Andreotti–Grauert

3.1. A real function φ of class C^∞, defined in a domain G of the space $\mathbf{C}^n$, is said to be *strongly p-convex*, if its Levi form

$$L(\varphi) = \sum_{1 \leq i, j \leq n} \frac{\partial^2 \varphi}{\partial z_i \, \partial \bar{z}_j} \, dz_i \, d\bar{z}_j$$

at each point of the domain G has at least $n - p + 1$ positive eigenvalues. A real function φ, defined on a complex space X, is said to be *strongly p-convex*, if for each sufficiently small open set $U \subset X$, realized as an analytic set in a domain G of the space $\mathbf{C}^n$, the function φ coincides with the restriction to U of a strongly p-convex function defined in the domain G. A real function φ on a complex space X is said to be *exhaustive* if for each real number c the set

$$B_c = \{x \in X: \varphi(x) < c\}$$

is relatively compact in X. The complex space X is said to be *strongly p-convex* if on X there exists an exhaustive function which is strongly p-convex away from a compact set.[11]

Theorem. Let X be a strongly p-convex complex space and $\mathcal{F}$ be a coherent analytic sheaf on X. Then the cohomology vector spaces $H^k(X; \mathcal{F})$ are finite-dimensional for $k \geq p$.[12]

See the proof in Andreotti and Grauert [1].[13]

Corollary. Let X be a strongly p-convex complex space and $\mathscr{F}$ be a coherent analytic sheaf on X. Then the homology vector spaces $H_k^c(X; \mathscr{F})$ are finite-dimensional for $k \geq p$.

Proof. By the corollary to Theorem 2.3 of Chapter 3, the topological vector spaces $H_k^c(X; \mathscr{F})$ are separated for $k \geq p - 1$. The assertion follows now from Theorem 2.3 of Chapter 3.

3.2. Theorem. Let X be a strongly p-convex complex space of finite dimension and $\mathscr{F}$ be a coherent analytic sheaf on X of homological codimension h. Then the homology vector spaces $H_k(X; \mathscr{F})$ are finite-dimensional for $k \leq h - p$.

Proof. We use the spectral sequence

$$E_{r,s}^2 = H^{-r}(X; \mathscr{H}_s(\mathscr{F})) \Rightarrow H_{r+s}(X; \mathscr{F})$$

(cf. Theorem 4.1 of Chapter 2). Since $\mathscr{H}_s(\mathscr{F}) = 0$ for $s < h$ and for $s > \dim X$ (cf. Theorem 1.1), the vector spaces $E_{r,s}^2$ are finite-dimensional for $r \leq -p$, and $E_{r,s}^2 = 0$ if $s < h$ or $s > \dim X$. Thus the theorem is proved.

Corollary. Let X be a strongly p-convex complex space of finite dimension and $\mathscr{F}$ be a coherent analytic sheaf on X of homological codimension h. Then the cohomology vector spaces $H_c^k(X; \mathscr{F})$ are finite-dimensional for $k \leq h - p$.[14]

Proof. By the corollary to Theorem 3.3 of Chapter 3 the topological vector spaces $H_c^k(X; \mathscr{F})$ are separated for $k \leq h - p + 1$. Hence the assertion follows from Theorem 3.3 of Chapter 3.

3.3. A real function φ of class C^∞, defined in a domain G of the space $\mathbf{C}^n$, is said to be *strongly q-concave*, if its Levi form $L(\varphi)$ at each point of the domain G has at least $n - q + 1$ negative eigenvalues. A real function φ, defined on a complex space X, is said to be *strongly q-concave*, if for each sufficiently small open set $U \subset X$, realized as an analytic set in a domain G of the space $\mathbf{C}^n$, the function φ coincides with the restriction to U of a strongly q-concave function, defined in the domain G. A complex space X is said to be *strongly q-concave*, if on X there exists a continuous exhaustive function, which is strongly q-concave away from a compact set.

Theorem. Let X be a strongly q-concave complex space and $\mathscr{F}$ be a coherent analytic sheaf on X. Then the cohomology vector spaces $H_c^k(X; \mathscr{F})$ are finite-dimensional for $k > q$.

See the proof in Andreotti and Kas [2].

Corollary. Let X be a strongly q-concave complex space and $\mathscr{F}$ be a coherent analytic sheaf on X. Then the homology vector spaces $H_k(X; \mathscr{F})$ are finite-dimensional for $k > q$.

Proof. By the corollary to Theorem 3.3 of Chapter 3, the topological vector spaces $H_k(X; \mathscr{F})$ are separated for $k \geq q$. Hence the assertion follows from Theorem 3.3 of Chapter 3.

3.4. Theorem. Let X be a strongly q-concave complex space of finite dimension and $\mathscr{F}$ be a coherent analytic sheaf on X of homological codimension h. Then the homology vector spaces $H_k{}^c(X; \mathscr{F})$ are finite-dimensional for $k < h - q$.

Proof. We use the spectral sequence

$$E_{r,s}^2 = H_{\bar{c}}^{-r}(X; \mathscr{H}_s(\mathscr{F})) \Rightarrow H_{r+s}^c(X; \mathscr{F})$$

(cf. Theorem 4.1 of Chapter 2). Since $\mathscr{H}_s(\mathscr{F}) = 0$ for $s < h$ and for $s > \dim X$ (cf. Theorem 1.1), the vector spaces $E_{r,s}^2$ are finite-dimensional for $r < -q$ and $E_{r,s}^2 = 0$, if $s < h$ or $s > \dim X$. Thus the theorem is proved.

Corollary. Let X be a strongly q-concave complex space of finite dimension and $\mathscr{F}$ be a coherent analytic sheaf on X of homological codimension h. Then the cohomology vector spaces $H^k(X; \mathscr{F})$ are finite-dimensional for $k < h - q$.[12]

Proof. By the corollary to Theorem 2.3 of Chapter 3, the topological vector spaces $H^k(X; \mathscr{F})$ are separated for $k \leq h - q$. Hence the assertion follows from Theorem 2.3 of Chapter 3.[15]

4. Lefschetz Theorem

4.1. Let M be an analytic set in a domain G of the space $\mathbf{C}^n$, defined as a ringed space by a coherent subsheaf of ideals $\mathscr{I}$ in $\mathcal{O}_G$ (cf. points 1.11.1 and 1.11.4 of Chapter 1). For each integer $p \geq 0$ the sheaf

$$\Omega_M^p = \Omega_G^p / (\mathscr{I}\Omega_G^p + d\mathscr{I} \wedge \Omega_G^{p-1}) | M$$

is called *the sheaf of germs of holomorphic forms* of degree p on M. The exterior differential $d: \Omega_G{}^p \to \Omega_G{}^{p+1}$ induces a homomorphism of sheaves

$$d: \ \Omega_M^p \longrightarrow \Omega_M^{p+1},$$

which is also called the *exterior differential*.

Let X be a complex space and (M_i) be a sufficiently fine covering of it by open sets. Then each M_i can be realized as an analytic set in a domain G_i of the space $\mathbf{C}^{n_i}$. For each pair of indices (i, j), there exists a holomorphic map $\varphi_{ij}: G_{ij} \to G_{ji}$ of an open neighborhood $G_{ij} \subset G_i$ of the set $M_{ij} = M_i \cap M_j$ into an open neighborhood $G_{ji} \subset G_j$ of this same set, inducing the identity isomorphism of the set M_{ij} as a ringed space onto itself. The map φ_{ij} induces over M_{ij} an isomorphism of analytic sheaves $\rho_{ji}: \Omega_{M_j}^p \to \Omega_{M_i}^p$, where ρ_{ii} is the identity isomorphism, and for each triple of indices (i, j, k) over M_{ijk} one has the equality $\rho_{ki} = \rho_{ji} \circ d$. From this it follows (cf. Serre [3, p. 376]) that on X there exists an analytic sheaf Ω_X^p, which is unique up to isomorphism, for which over each M_i one has an isomorphism ζ_j : $\Omega_{M_i}^p \to \Omega_{M_X}^p$, such that for each pair of indices (i, j), over the intersection M_{ij} one has $\zeta_j = \zeta_i \circ \rho_{ji}$. The sheaf Ω_X^p is called *the sheaf of germs of holomorphic forms* of degree p on X.[16]

For each pair of indices (i, j), over the intersection M_{ij} one has $d \circ \rho_{ji} = \rho_{ji} \circ d$. Consequently, there exists a homomorphism of sheaves

$$d: \ \Omega_X^p \longrightarrow \Omega_X^{p+1},$$

which is called *the exterior differential* and over each set M_i it satisfies the equation $d \circ \zeta_j = \zeta_i \circ d$. Since $d \circ d = 0$, obviously, the sheaves Ω_X^p ($p = 0$, 1, ...), together with their differential d, form a *complex* $\Omega_X^\cdot$.[17]

4.2. Let X be a complex space which is countable at infinity, and $\mathscr{S}_X^p$ for each integer $p \geq 0$ be the sheaf of localized singular differentiable cochains on X with values in $\mathbf{C}$ (cf. Godement [1, p. 184]). The sheaves $\mathscr{S}_X^p$ ($p = 0, 1, ...$) together with the coboundary operator

$$\delta: \ \mathscr{S}_X^p \longrightarrow \mathscr{S}_X^{p+1}$$

form a complex $\mathscr{S}_X^\cdot$. By Stokes' formula there exists a natural cochain map of complexes $\Omega_X^\cdot \to \mathscr{S}_X^\cdot$, for which the diagram

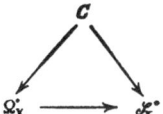

is commutative. Passing to hypercohomology with an arbitrary family of supports Φ, we get a commutative diagram

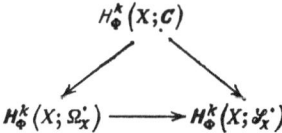

Since the complex $\mathscr{S}^{\cdot}_X$ is a flabby resolution of the simple sheaf C (cf. Godement [1, p. 185]), there is a natural isomorphism

$$H^k_\Phi(X; \mathscr{S}^{\cdot}_X) = H^k_\Phi(X; C).$$

From this we get directly the *Bloom–Herrera decomposition*

$$H^k_\Phi(X; \Omega^{\cdot}_X) = H^k_\Phi(X; C) \oplus A^k_\Phi$$

(cf. Bloom and Herrera [1, p. 288]).[18]

4.3. Let X be an arbitrary complex space. We set

$$\rho(X) = \inf_p \text{codh } \Omega^p_X.$$

Since codh $\Omega_X{}^p \leq \dim X$ for each p, one has $\rho(X) \leq \dim X$. If X is a complex manifold of dimension n, then $\rho(X) = n$.

Theorem. If the complex space X is holomorphically complete, then[19]

$$H^k_c(X; C) = 0 \quad \text{for} \quad k < \rho(X).$$

Proof. We use the spectral sequence

$$E^2_{p,q} = H^{-p}(X; \mathscr{H}_q(\mathscr{F})) \Rightarrow H_{p+q}(X; \mathscr{F})$$

(cf. Theorem 4.1 of Chapter 2). If $\mathscr{F}$ is a coherent analytic sheaf on X, then the sheaves of germs of homology $\mathscr{H}_q(\mathscr{F})$ are also coherent. By Cartan's Theorem (B) (cf. point 1.13 of Chapter 1) $E_{p,q}{}^2 = 0$ for $p \neq 0$, i.e., the spectral sequence degenerates. We get a natural isomorphism

$$H_k(X; \mathcal{F}) = \Gamma(X; \mathcal{H}_k(\mathcal{F})).$$

Since $\mathcal{H}_k(\mathcal{F}) = 0$ for $k < \mathrm{codh}\, \mathcal{F}$, one has

$$H_k(X; \mathcal{F}) = 0 \quad \text{for} \quad k < \mathrm{codh}\, \mathcal{F}.$$

Now we set $\mathcal{F} = \Omega_X^p$ and we apply Theorem 3.3 of Chapter 3. We get

$$H_c^k(X; \Omega_X^p) = 0 \quad \text{for} \quad k < \mathrm{codh}\, \Omega_X^p.$$

We use the spectral sequence

$$E_2^{p,\,q} = H^p(H_c^q(X; \Omega_X^{\cdot})) \Rightarrow H_c^{p+q}(X; \Omega_X^{\cdot})$$

(cf., e.g., Grothendieck [2, p. 47]). From this we get that

$$H_c^k(X; \Omega_X^{\cdot}) = 0 \quad \text{for} \quad k < \rho(X).$$

The assertion of the theorem now follows in view of the Bloom–Herrera decomposition (cf. point 4.2). The theorem is proved.

Corollary. Let X be an arbitrary complex space, U be a holomorphically complete open set in X, $\rho = \rho(U)$. Then the restriction map

$$H_c^k(X; \boldsymbol{C}) \to H_c^k(X \setminus U; \boldsymbol{C})$$

is bijective for $k < \rho - 1$ and injective for $k = \rho - 1$.[20]

Proof. We use the cohomology exact sequence connected with the closed subset $X \setminus U$,

$$\ldots \to H_c^k(U; \boldsymbol{C}) \to H_c^k(X, \boldsymbol{C}) \to H_r^k(X \setminus U; \boldsymbol{C}) \to H_c^{k+1}(U; \boldsymbol{C}) \to \ldots$$

(cf. point 1.8.1 of Chapter 1). In view of Theorem 4.3 the assertion follows from this directly.

5. Theory of Hyperfunctions

5.1. Let M be an *analytic set* in a domain G of the space $\boldsymbol{R}^n$, i.e., the set of common zeros of a finite number of real analytic functions on G. The set M becomes a ringed space by endowing it with the structure sheaf of complex algebras

$$\mathscr{A}_M = \mathscr{A}_G / \mathscr{I} \mid M$$

(cf. point 1.11.1 of Chapter 1), where $\mathscr{A}_G$ is the sheaf of germs of analytic functions on G, assuming complex values, and $\mathscr{I}$ is the coherent subsheaf of ideals in $\mathscr{A}_G$, whose set of zeros coincides with M.

By a *real analytic space* is meant a separated topological space X, endowed with a structure sheaf of complex algebras $\mathscr{A}_X$ (i.e., which is a ringed space) such that each point in X has an open neighborhood, isomorphic as a ringed space with an analytic set in a domain of the space $\mathbf{R}^n$.[21]

The complex space $\widetilde{X}$ is called a *complexification* of the real analytic space X, if the space X is isomorphic as a ringed space to a closed set in the space $\widetilde{X}$, endowed with the structure sheaf induced by the sheaf $\mathcal{O}_{\widetilde{X}}$. By the Whitney–Bruhat theorem such a complexification $\widetilde{X}$ exists for any real analytic space X which is countable at infinity (cf. Whitney and Bruhat [1] and Hironaka [1]).[22]

5.2. Lemma. Let X be a real analytic space which is countable at infinity, $\widetilde{X}$ be a complexification of it, and $\mathscr{F}$ be a coherent analytic sheaf on $\widetilde{X}$. Then for any sufficiently small open set $\widetilde{U} \subset \widetilde{X}$

$$H_k^X(\widetilde{U}; \mathscr{F}) = 0 \quad \text{for} \quad k \neq 0.$$

In particular,

$$\mathscr{H}_k^X(\mathscr{F}) = 0 \quad \text{for} \quad k \neq 0.$$

Proof. One can assume that X is a closed set in the holomorphically complete complex space $\widetilde{X}$ (cf. Grauert [2]).[23] Analogously, for each compact set $S \subset X$ in the space $\widetilde{X}$ there exists a fundamental system of open neighborhoods which are holomorphically convex relative to $\widetilde{X}$. By Theorem 3.9 of Chapter 4,

$$H_k^S(\widetilde{X}; \mathscr{F}) = 0 \quad \text{for} \quad k \neq 0,$$

and the topological vector space $H_0^S(\widetilde{X}; \mathscr{F})$ is separated and naturally isomorphic to the strong dual to the topological vectors space $\Gamma(S; \mathscr{F})$. In particular, this is true for the compact sets ∂U and U, where $U = \tilde{U} \cap X$. On the other hand, there is the exact sequence for homology groups

$$\ldots \to H_k^{\partial U}(\tilde{X}; \mathscr{F}) \to H_k^{\bar{U}}(\tilde{X}; \mathscr{F}) \to H_k^X(\tilde{U}; \mathscr{F}) \to H_{k-1}^{\partial U}(\tilde{X}; \mathscr{F}) \to \ldots$$

(cf. point 2.6 of Chapter 2). Since the restriction map

$$\Gamma(\bar{U}; \mathscr{F}) \to \Gamma(\partial U; \mathscr{F})$$

has everywhere dense image (cf., e.g., Gunning and Rossi [1, p. 308]), the dual map

$$H_0^{\partial U}(\tilde{X}; \mathscr{F}) \to H_0^{\bar{U}}(\tilde{X}; \mathscr{F})$$

is injective. The assertion follows directly from this. The lemma is proved.[24]

5.3. Theorem. Let X be a complex space, $\mathscr{F}$ be an analytic sheaf on X, and S and T be closed sets in X. Then there exists a spectral sequence

$$E_{p,q}^2 = H_S^{-p}(X; \mathscr{H}_q^T(\mathscr{F})) \Rightarrow H_{p+q}^{S \cap T}(X; \mathscr{F}).$$

Proof. Let

$$0 \to \mathscr{F} \to \mathscr{L}^0 \to \mathscr{L}^1 \to \ldots$$

be a locally fine resolution of the analytic sheaf $\mathscr{F}$. We choose an injective resolution in the sense of Cartan–Eilenberg

$$0 \to \mathscr{H}_0^T(\mathscr{L}^\bullet) \to \mathscr{L}_{0,\bullet} \to \mathscr{L}_{-1,\bullet} \to \ldots$$

of the complex $\mathscr{H}_0^T(\mathscr{L}^\bullet)$ (cf. Cartan and Eilenberg [1, Chapter XVII, point 1, p. 437]) and we consider the double complex

$$K_{p,q} = \Gamma_S(X; \mathscr{L}_{p,q}).$$

For its first spectral sequence

$$'E_{p,q}^1 = \Gamma_S(X; H_q \mathscr{L}_{p,\bullet}).$$

Since for fixed q we have an injective resolution

$$0 \to H_q \mathcal{H}_0^T (\mathcal{L}^{\cdot}) \to H_q \mathcal{L}_{0,\cdot} \to H_q \mathcal{L}_{-1,\cdot} \to \cdots$$

of the sheaf $H_q \mathcal{H}_0^T(\mathcal{L}^{\cdot}) = \mathcal{H}_q^T(\mathcal{F})$ (cf. Theorem 2.4 and Proposition 2.5 of Chapter 2), so

$$'E_{p,q}^2 = H_S^{-p}(X; \ \mathcal{H}_q^T(\mathcal{F})).$$

On the other hand, for the second spectral sequence

$$^{\bullet}E_{p,q}^1 = H_S^{-q}(X; \ \mathcal{H}_0^T(\mathcal{L}^p)).$$

By Lemma 2.1 of Chapter 2, the sheaf $\mathcal{H}_0^T(\mathcal{L}^p) = {}_T\mathcal{H}_0(\mathcal{L}^p)$ is flabby, so

$$"E_{p,q}^1 = 0 \quad \text{for} \quad q \neq 0,$$
$$"E_{p,0}^1 = \Gamma_{S \cap T}(X; \ \mathcal{H}_0(\mathcal{L}^p)).$$

Thus the second spectral sequence degenerates:

$$"E_{p,q}^2 = 0 \quad \text{for} \quad q \neq 0;$$
$$"E_{p,0}^2 = H_p^{S \cap T}(X; \ \mathcal{F}).$$

The theorem is proved.

5.4. Theorem. Let X be a real analytic space which is countable at infinity, $\widetilde{X}$ be a complexification of it, and $\mathcal{F}$ be a coherent analytic sheaf on $\widetilde{X}$. Then for any open set $\widetilde{U} \subset \widetilde{X}$

$$H_k^X(\widetilde{U}; \ \mathcal{F}) = 0 \quad \text{for} \quad k \neq 0;$$
$$H_0^X(\widetilde{U}; \ \mathcal{F}) = \Gamma(\widetilde{U}; \ \mathcal{H}_0^X(\mathcal{F})).$$

Proof. We use the spectral sequence

$$E_{p,q}^2 = H^{-p}(\widetilde{U}; \ \mathcal{H}_q^X(\mathcal{F})) \Rightarrow H_{p+q}^X(\widetilde{U}; \ \mathcal{F})$$

(cf. Theorem 5.3). By Lemma 5.2, $\mathcal{H}_q^X(\mathcal{F}) = 0$ for $q \neq 0$; consequently, the spectral sequence degenerates. The assertion follows from this. The theorem is proved.[25]

5.5. Theorem. Let X be a real analytic space which is countable at infinity, $\widetilde{X}$ be its complexification, and $\mathcal{F}$ be a coherent analytic sheaf on $\widetilde{X}$. Then $\mathcal{H}_0^X(\mathcal{F})$ is a flabby sheaf on $\widetilde{X}$.

Proof. Let $\tilde{U}$ be an arbitrary open set in $\tilde{X}$ and $U = \tilde{U} \cap X$. Then there is an exact sequence of homology groups

$$\ldots \to H_0^{X \setminus U}(\tilde{X};\ \mathcal{F}) \to H_0^X(\tilde{X};\ \mathcal{F}) \to H_0^X(\tilde{U};\ \mathcal{F}) \to 0$$

(cf. point 2.6 of Chapter 2). In particular, the restriction map

$$H_0^X(\tilde{X};\ \mathcal{F}) \to H_0^X(\tilde{U};\ \mathcal{F})$$

is surjective. From this, in view of Theorem 5.4, the assertion follows. The theorem is proved.

5.6. Let X be a real analytic space which is countable at infinity, $\tilde{X}$ be its complexification. Then the sheaf

$$\mathcal{B}_X = \mathcal{H}_0^X(\mathcal{O}_{\tilde{X}}) \,|\, X$$

is called *the sheaf of germs of hyperfunctions* on X. By Theorem 5.5 the sheaf $\mathcal{B}_X$ is flabby.[26]

Sections of the sheaf $\mathcal{B}_X$ are called *hyperfunctions*. By Theorem 5.4, for each open set $U \subset X$ there is a natural isomorphism

$$\Gamma(U;\ \mathcal{B}_X) = H_0^X(\tilde{U};\ \mathcal{O}_{\tilde{X}}),$$

where $\tilde{U}$ is an arbitrary open set in $\tilde{X}$, for which $U = \tilde{U} \cap X$; in other words, hyperfunctions are zero-dimensional homology classes of the structure sheaf $\mathcal{O}_{\tilde{X}}$ with supports in X.[27]

5.7. Theorem. Let M be an analytic set in a domain G of the space $\mathbf{R}^n$, defined as the set of common zeros of the analytic functions $f_1, \ldots, f_p$ in G. Then there is a natural isomorphism of vector spaces

$$\Gamma(M;\ \mathcal{B}_M) = \{\alpha \in \Gamma(G;\ \mathcal{B}_G): f_i \alpha = 0 \ (1 \leqslant i \leqslant p)\}.$$

Proof. We define a sheaf of complex algebras $\mathcal{A}_M = \mathcal{A}_G / \mathcal{J} \,|\, M$ on M, where $\mathcal{A}_G$ is the sheaf of germs of analytic functions on G, and $\mathcal{J}$ is the subsheaf of ideals in $\mathcal{A}_G$, generated by the functions $f_1, \ldots, f_p$. We choose a domain $\tilde{G} \subset \mathbf{C}^n$, for which $G = \tilde{G} \cap \mathbf{R}^n$, and we assume that each function $f_1, \ldots, f_p$ is extended to a holomorphic function on $\tilde{G}$. Let $\tilde{M}$ be an analytic set in the domain $\tilde{G}$, which coincides with the set of common zeros of the holomorphic functions $f_1, \ldots, f_p$ in $\tilde{G}$. We define the sheaf of complex algebras $\mathcal{O}_{\tilde{M}} = \mathcal{O}_{\tilde{G}}/\mathcal{Y} | M$ on $\tilde{M}$, where $\mathcal{O}_{\tilde{G}}$ is the sheaf of germs of holomorphic functions on $\tilde{G}$, and $\mathcal{Y}$ is the subsheaf of ideals in $\mathcal{O}_{\tilde{G}}$, generated by the functions $f_1, \ldots, f_p$. Since the sheaf $\mathcal{O}_{\tilde{M}}$ induces the sheaf $\mathscr{A}_M$ on M, one has that $\tilde{M}$ is a complexification for M. In the domain G there is defined the sheaf of germs of hyperfunctions $\mathscr{B}_G = \mathscr{H}_0{}^G(\mathcal{O}_{\tilde{G}}) \,|\, G$, and on the set M, the sheaf of germs of hyperfunctions

$$\mathscr{B}_M = \mathscr{H}_0^M\,(\mathcal{O}_{\tilde{M}})\,|\,M = \mathscr{H}_0^G\,(\mathcal{O}_{\tilde{G}}/\mathcal{Y})\,|\,M.$$

From the exact sequence

$$\mathcal{O}_{\tilde{G}}^p \xrightarrow{\;(f_1,\,\ldots,\,f_p)\;} \mathcal{O}_{\tilde{G}} \to \mathcal{O}_{\tilde{G}}/\mathcal{Y} \to 0$$

by virtue of the exactness of the functor $\mathscr{F} \mapsto \mathscr{H}_0{}^G(\mathscr{F})$ (cf. Lemma 5.2) we get the exact sequence

$$0 \to \mathscr{H}_0^G\,(\mathcal{O}_{\tilde{G}}/\mathcal{Y}) \to \mathscr{H}_0^G\,(\mathcal{O}_{\tilde{G}}) \xrightarrow{\;(f_1,\,\ldots,\,f_p)\;} \mathscr{H}_0^G\,(\mathcal{O}_{\tilde{G}}^p).$$

In other words, on M the sequence of sheaves

$$0 \to \mathscr{B}_M \to \mathscr{B}_G \xrightarrow{\;(f_1,\,\ldots,\,f_p)\;} \mathscr{B}_G^p$$

is exact. The theorem is proved.

5.8. **Theorem.** On the real analytic space X, which is countable at infinity, suppose there are given analytic functions $f_1, \ldots, f_p \in \Gamma(X; \mathscr{A}_X)$ and hyperfunctions $\alpha_1, \ldots, \alpha_p \in \Gamma_S(X; \mathscr{B}_X)$ with supports in the closed set $S \subset X$. Then in order that there exist a hyperfunction $\alpha \in \Gamma_S(X; \mathscr{B}_X)$, which satisfies the equalities

$$f_i \alpha = \alpha_i \qquad (i = 1, \ldots, p),$$

it is necessary and sufficient that the following condition hold: for each point $x \in S$, any relation of the form

$$\theta_1 f_1 + \ldots + \theta_p f_p = 0,$$

where $\theta_1, \ldots, \theta_p$ are analytic functions in a neighborhood of the point x, implies the relation[28]

$$\theta_1\alpha_1 + \ldots + \theta_p\alpha_p = 0 .$$

Proof. Let $\tilde{X}$ be a complexification of the space X, to which all the functions $f_1, \ldots, f_p$ admit complex analytic continuation. The functions $f_1, \ldots, f_p$ then define a homomorphism of analytic sheaves

$$\mathcal{O}_{\tilde{X}}^p \xrightarrow{(f_1, \ldots, f_p)} \mathcal{O}_{\tilde{X}},$$

whose kernel we denote by $\mathcal{R}$. We get an exact sequence of homology groups

$$H_0^S(\tilde{X};\, \mathcal{O}_{\tilde{X}}) \xrightarrow{(f_1, \ldots, f_p)} H_0^S(\tilde{X};\, \mathcal{O}_{\tilde{X}}^p) \to H_0^S(\tilde{X};\, \mathcal{R}) \to 0.$$

Since the natural map

$$H_0^S(\tilde{X};\, \mathcal{R}) \to \prod_{x \in S} \mathcal{H}_0^S(\mathcal{R})_x$$

is injective, we get an exact sequence

$$\Gamma_S(X;\, \mathcal{B}_X) \xrightarrow{(f_1, \ldots, f_p)} \Gamma_S(X;\, \mathcal{B}_X)^p \to \prod_{x \in S} \mathcal{H}_0^S(\mathcal{R})_x.$$

In a neighborhood of each point $x \in S$ there is defined an epimorphism of sheaves $\mathcal{O}_{\tilde{X}}^q \to \mathcal{R}$. Consequently, the map

$$\mathcal{H}_0^S(\mathcal{R}) \to \mathcal{H}_0^S(\mathcal{O}_{\tilde{X}}^q)$$

is injective in a neighborhood of the point x, and the sequence

$$\mathcal{O}_{\tilde{X}}^q \to \mathcal{O}_{\tilde{X}}^p \xrightarrow{(f_1, \ldots, f_p)} \mathcal{O}_{\tilde{X}}$$

is exact. Thus there is an exact sequence

$$\Gamma_S(X;\, \mathcal{B}_X) \xrightarrow{(f_1, \ldots, f_p)} \Gamma_S(X;\, \mathcal{B}_X)^p \to \prod_{x \in S} {}_S\mathcal{B}_{X,x}^q.$$

The theorem is proved.

5.9. Theorem. The sheaf of germs of hyperfunctions $\mathcal{B}_X$ on the real analytic space X, which is countable at infinity, is an injective $\mathcal{A}_X$-module.[29]

Proof. Since the injectivity of a sheaf is a local property, one can assume that $X = M$ is an analytic set in a domain G of the space $\mathbf{R}^n$. It suffices to prove that the sheaf of germs of hyperfunctions $\mathscr{B}_M$ is an injective $\mathscr{A}_M$-module. Since $\mathscr{A}_M = \mathscr{A}_G / \mathscr{I} M$ and $\mathscr{B}_M = {}_{\mathscr{I}}\mathscr{B}_G$ (cf. Theorem 5.7), by Lemma 2.6 of Chapter 1 it suffices to prove that the sheaf $\mathscr{B}_G$ is an injective $\mathscr{A}_G$-module. Let $\tilde{G}$ be a domain in $\mathbf{C}^n$, which is a complexification of the domain G. We extend the sheaf $\mathscr{B}_G$ trivially to the domain $\tilde{G}$ and we shall consider it as an $\mathcal{O}_{\tilde{G}}$-module. Since $\mathcal{O}_{\tilde{G}} | G = \mathscr{A}_G$, it suffices to prove that $\mathscr{B}_G$ is an injective $\mathcal{O}_{\tilde{G}}$-module. But this follows from Theorem 2.8 of Chapter 1, since $\mathscr{B}_G$ is a flabby sheaf (cf. Theorem 5.5), for which the stalks ${}_S\mathscr{B}_{G,x}$ are injective $\mathscr{A}_{G,x}$-modules (cf. Theorem 5.8). The theorem is proved.[30]

5.10. Theorem. Let X be a complex space, $\mathscr{F}$ be an analytic sheaf on X, and S and T be closed sets in X. Then there exists a spectral sequence

$$E_{p,q}^2 = \operatorname{Ext}_{\mathcal{O}_X, S}^{-p}(X; \mathscr{F}, \mathscr{H}_q^T(\mathcal{O}_X)) \Rightarrow H_{p+q}^{S \cap T}(X; \mathscr{F}).$$

Proof. We choose a locally fine resolution

$$0 \to \mathcal{O}_X \to \mathscr{L}^0 \to \mathscr{L}^1 \to \ldots$$

of the sheaf $\mathcal{O}_X$, consisting of flat $\mathcal{O}_X$-modules. We construct an injective resolution in the sense of Cartan–Eilenberg

$$0 \to \mathscr{H}_0^T(\mathscr{L}^\cdot) \to \mathscr{L}_{0,.} \to \mathscr{L}_{-1,.} \to \ldots$$

of the complex $\mathscr{H}_0^T(\mathscr{L}^\cdot)$ (cf. Cartan and Eilenberg [1, Chapter XVII, point 1, p. 437]). We consider the double complex

$$K_{p,q} = \operatorname{Hom}_{\mathcal{O}_X, S}(X; \mathscr{F}, \mathscr{L}_{p,q}),$$

where $\mathscr{F}$ is an arbitrary analytic sheaf on X. For its first spectral sequence

$$'E_{p,q}^1 = \operatorname{Hom}_{\mathcal{O}_X, S}(X; \mathscr{F}, H_q \mathscr{L}_{p,.}).$$

Since for fixed q we get an injective resolution

$$0 \to H_q \mathscr{H}_0^T(\mathscr{L}^\cdot) \to H_q \mathscr{L}_{0,.} \to H_q \mathscr{L}_{-1,.} \to \ldots$$

of the sheaf $H_q \mathscr{H}_0^T(\mathscr{L}^\cdot) = \mathscr{H}_q^T(\mathcal{O}_X)$ (cf. Theorem 2.4 and Proposition 2.5 of Chapter 2), we have

$$'E^2_{p,q} = \text{Ext}^{-p}_{\mathcal{O}_X, \, S}(X; \, \mathcal{F}, \, \mathcal{H}^T_q(\mathcal{O}_X)).$$

For the second spectral sequence of the double complex $(K_{p,q})$, we have

$$''E^1_{p,q} = \text{Ext}^{-q}_{\mathcal{O}_X, \, S}(X; \, \mathcal{F}, \, \mathcal{H}^T_0(\mathcal{L}^p)).$$

By Lemma 3.2 of Chapter 2, the $\mathcal{O}_X$-module $\mathcal{H}_0^T(\mathcal{L}^p) = {}_T\mathcal{H}_0(\mathcal{L}^p)$ is injective (cf. the corollary to Lemma 2.6 of Chapter 1), so

$$''E^1_{p,q} = 0 \quad \text{for} \quad q \neq 0,$$
$$''E^1_{p,0} = \text{Hom}_{\mathcal{O}_X, \, S \cap T}(X; \, \mathcal{F}, \, \mathcal{H}_0(\mathcal{L}^p)).$$

Consequently, the second spectral sequence degenerates:

$$''E^2_{p,q} = 0 \quad \text{for} \quad q \neq 0,$$
$$''E^2_{p,0} = H_p \text{Hom}_{\mathcal{O}_X, \, S \cap T}(X; \, \mathcal{F}, \, \mathcal{H}_0(\mathcal{L}^{\boldsymbol{\cdot}})).$$

The assertion now follows from Theorem 3.1 of Chapter 2. The theorem is proved.

5.11. Theorem. Let X be a real analytic space which is countable at infinity, $\tilde{X}$ be its complexification, and $\mathcal{F}$ be a coherent analytic sheaf on $\tilde{X}$. Then for any open set $\tilde{U} \subset \tilde{X}$ there is a natural isomorphism

$$H^X_0(\tilde{U}; \, \mathcal{F}) = \text{Hom}_{\mathcal{A}_X}(U; \, \mathcal{F}, \, \mathcal{B}_X),$$

where $U = \tilde{U} \cap X$. In particular,[31]

$$\mathcal{H}^X_0(\mathcal{F}) \, | \, X = \mathcal{H} \text{om}_{\mathcal{A}_X}(\mathcal{F}, \, \mathcal{B}_X).$$

Proof. We use the spectral sequence

$$E^2_{p,q} = \text{Ext}^{-p}_{\mathcal{O}_{\tilde{X}}}(\tilde{U}; \, \mathcal{F}, \, \mathcal{H}^X_q(\mathcal{O}_{\tilde{X}})) \Rightarrow H^X_{p+q}(\tilde{U}; \, \mathcal{F})$$

(cf. Theorem 5.10). Since by Lemma 5.2, $\mathcal{H}_q^X(\mathcal{O}_{\tilde{X}}) = 0$ for $q \neq 0$, this spectral sequence degenerates. The assertion follows directly from this. The theorem is proved.

5.12. Remarks.

5.12.1. In the notation of Theorem 5.7, we consider the coherent analytic sheaf $\mathscr{F} = \mathcal{O}_{\tilde{G}}/\mathscr{Y}$ on $\tilde{G}$. By Theorem 5.11 we get a natural isomorphism

$$H_0^G(\bar{G};\, \mathcal{O}_{\tilde{G}}/\mathscr{Y}) = \operatorname{Hom}_{\mathcal{A}_G}(G;\, \mathcal{A}_G/\mathscr{I},\, \mathscr{B}_G).$$

Since

$$\Gamma(M;\, \mathscr{B}_M) = H_0^G(\bar{G};\, \mathcal{O}_{\tilde{G}}/\mathscr{Y}),$$

the assertion of Theorem 5.7 follows from this.

5.12.2. In the notation of Theorem 5.8 we consider the exact sequence of sheaves

$$0 \longrightarrow \mathscr{R} \longrightarrow \mathcal{O}_{\tilde{X}}^p \xrightarrow{(f_1,\ldots,\, f_p)} \mathcal{O}_{\tilde{X}}.$$

We get an exact sequence of homology groups

$$H_0^X(\bar{X};\, \mathcal{O}_{\tilde{X}}) \xrightarrow{(f_1,\ldots,\, f_p)} H_0^X(\bar{X};\, \mathcal{O}_{\tilde{X}}^p/\mathscr{R}) \longrightarrow 0.$$

By Theorem 5.11 this means that the map

$$\Gamma(X;\, \mathscr{B}_X) \xrightarrow{(f_1,\ldots,\, f_p)} \operatorname{Hom}_{\mathcal{A}_X}(X;\, \mathcal{O}_{\tilde{X}}^p/\mathscr{R},\, \mathscr{B}_X)$$

is surjective. The assertion of Theorem 5.8 follows from this.

NOTES

Chapter 1

1. Sheaves were first introduced into topology by Leray [1], as well as cohomology groups with coefficients in sheaves and spectral sequences. Definition 1.1 of a sheaf of Abelian groups was first published in the memoirs of the Cartan seminar [2]. In this brief account of the general theory of sheaves, we follow Serre's paper [3] and Godement's book [1].

2. The definition of cohomology groups with coefficients in sheaves as right derived functors of the functor of global sections was first introduced by Grothendieck [2].

3. Apparently the first systematic account of the theory of sheaves and cohomology with coefficients in sheaves is contained in the mimeographed proceedings of the Cartan seminar [2]. Cohomology groups with coefficients in sheaves are defined there with the help of fine resolutions. Flabby and soft sheaves, canonical resolutions, and the definition of cohomology groups with coefficients in sheaves with their help were first introduced by Godement [1].

4. Aleksandrov [1, 2] associated with an arbitrary finite closed covering $\mathfrak{u} = (U_i)$ of a compact topological space X, an abstract simplicial complex $N(\mathfrak{u})$, which he called the *nerve* of the covering $\mathfrak{u}$. The vertices of the complex $N(\mathfrak{u})$ are the sets U_i, and the k-dimensional simplices are those collections of sets $(U_{i_0}, ..., U_{i_k})$ such that the intersection $U_{i_0} \cap \cdots \cap U_{i_k}$ is nonempty. P. S. Aleksandrov actually defined the cohomology groups of the space X as the projective limit of the homology groups of the complexes $N(\mathfrak{u})$ with respect to the

filtered set of all (finite) coverings of this space. An essential point: although P. S. Aleksandrov considered only compact spaces and finite coverings of them, his construction carries over without any special difficulty to arbitrary topological spaces and arbitrary (open or closed) coverings of them, which was later done by Čech [E. Čech, "Théorie générale de l'homologie dans un espace quelconque," *Fundam. Math.*, **19**, 149–183 (1932)]. Leray [1] and Cartan [4] used a dual construction (an analogous construction was considered earlier by Aleksandrov [3]) to define cohomology groups with coefficients in sheaves (cf. point 1.9). This briefly is the history of this question; however, in the literature (and some-times even in that of our country), the cohomology mentioned is incorrectly called Čech cohomology.

5.   The functors $\text{Ext}_{\mathcal{O}}^k(X; \mathscr{F}, \mathscr{G})$ were introduced by Grothendieck [1, 2] in connection with his generalization of the Serre duality theorem to algebraic varieties (cf. Hartshorne [2]).

6.   The concept of ringed space was first introduced by Cartan [3, 5], and Serre [4] also introduced the concept of (reduced) complex space. A more general definition of complex space was given by Grauert [4] (for references see Malgrange [6] and Onishchik [1]).

7.   The concept of coherent analytic sheaf was first introduced by Cartan [1]. Cartan's theorem on the coherence of the sheaf of ideals of an analytic set is also proved there (cf. Cartan [1, 3]). In connection with Oka's theorem, see Oka [1] and Cartan [3].

8.   The concept of holomorphically complete complex manifold was first introduced by Stein [1]. Holomorphically complete complex spaces are hence often called Stein spaces; however, this concept was only definitively formulated in Grauert [1]. The theory of coherent analytic sheaves on Stein manifolds was developed in the proceedings of the Cartan seminar [3] and, in particular, Cartan's fundamental theorems (A) and (B) were first proved there (cf. also Cartan [4]).

9.   For holomorphically complete complex manifolds, theorems (A) and (B) of Cartan were first proved in the proceedings of the Cartan seminar [3] (cf. also Cartan [4] and Malgrange [2]). For reduced complex spaces proofs were published by Andreotti and Vesentini [1], Gunning and Rossi [1], and Siu [1]; for general complex spaces, by Grauert [4]. Grauert and Remmert [2] and Onishchik [1] are specially devoted to the theory of holomorphically complete complex spaces.

10.   Theorem 1.15.1 on the coherence of direct images was first proved
       by Grauert [4]. Subsequently, simpler proofs were given by Knorr
       [1], Forster and Knorr [1], and Kiehl and Verdier [1].

11.   The study of injective sheaves of modules is of considerable interest
       in connection with the fact that injective resolutions are used for cal-
       culating the derived functors on the category of sheaves of modules
       and, in particular, for calculating cohomology groups with
       coefficients in sheaves (cf. points 1.6 and 1.7). For us in this book
       the problem of finding criteria for the injectivity of analytic sheaves
       is quite important, since the proofs of some of the basic properties
       of the homology groups of analytic sheaves are based on these cri-
       teria (cf. Chapter 2).
           The importance of the concept of injectivity of a sheaf in com-
       parison with the corresponding concept for ordinary modules is
       clear from the following remarks. Let θ be a sheaf of commutative
       rings with units on the topological space X and $\mathscr{F}$ be an injective θ-
       module. For each open set $U \subset X$ we define sections $f_U \in \Gamma(U; \mathscr{F})$
       and $\varphi_U \in \Gamma(U; \theta)$. Let us assume that for any finite number of open
       sets $U_1, ..., U_k$ in X at each point $x \in U_1 \cap ... \cap U_k$ one has

$$\alpha_1 f_{U_1, x} + ... + \alpha_k f_{U_k, x} = 0,$$

       where $\alpha_1, ..., \alpha_k \in \theta_x$ are arbitrary germs, satisfying the relation

$$\alpha_1 \varphi_{U_1, x} + ... + \alpha_k \varphi_{U_k, x} = 0.$$

       Then there exists a section $f \in \Gamma(X; \mathscr{F})$, such that $\varphi_U(f | U) = f_U$ for
       each U.

12.   Thus, the basic content of points 2.2 and 2.3 is the almost word for
       word carrying over of the theory of injective hulls of Baire (cf.
       Eckman and Schopf [1] and MacLane [1, pp. 137–139]) to sheaves
       of modules over sheaves of rings.

13.   The basic meaning of the hypotheses of Lemma 2.7 is the following:
       the point z_k is chosen arbitrarily in U_{k-1}; the family of holomorphic
       functions F_k is defined arbitrarily in a neighborhood of z_k; the open
       neighborhood U_k of the point z_k should be chosen sufficiently small
       depending on the family F_k. The choice is made by induction,
       separately at each step.
           For $z_1 = z_2 = ...$ from Lemma 2.7 we get the theorem that the
       ring of germs of holomorphic functions is Noetherian (cf. Gunning
       and Rossi [1, p. 95]).

14.   The following assertion follows from Lemma 2.7:

Let $\mathcal{O}$ be the sheaf of germs of holomorphic functions in the domain G of the space $\mathbf{C}^n$ and $\mathcal{F}$ be an arbitrary analytic subsheaf of the sheaf $\mathcal{O}^k$. Then the sheaf $\mathcal{F}$ is coherent on some open everywhere dense set $U \subset G$.

On the other hand, for any anywhere dense open set $U \subset G$ there exists a subsheaf of ideals $\mathcal{F} \subset \mathcal{O}$, which is coherent on U and is not coherent in a neighborhood of each point of $G \setminus U$ (it suffices to set $\mathcal{F} = \mathcal{O}_U$).

15.   The question of a characterization of injective sheaves of modules over arbitrary sheaves of rings is difficult and still not completely solved. An exhaustive solution can be obtained (analogously to Theorem 2.8) for Euclidean sheaves of rings (and quasi-Euclidean sheaves of rings; for example, for structure sheaves of rings of complex spaces and algebraic varieties). In particular, for sheaves of Abelian groups (i.e., $\mathbf{Z}$-modules) the following result holds:

A sheaf of Abelian groups $\mathcal{F}$ on a topological space X is injective (as a $\mathbf{Z}$-module) if and only if the following conditions hold:

a) the sheaf $\mathcal{F}$ is flabby;

b) for any closed set $S \subset X$, for each point $x \in X$ the stalk of the sheaf $s\mathcal{F}$ (cf. point 2.6) over x is a divisible (i.e., complete) Abelian group.

The question of a characterization of injective sheaves of Abelian groups was considered by Dobbs [D. E. Dobbs, *Can. J. Math.*, **29**, No. 5, 1031–1039 (1977); **32**, No. 6, 1522 (1980)] and Banaschewski [B. Banaschewski, *Can. J. Math.*, **32**, No. 6, 1518–1521 (1980)]. Dobbs asserted that a sheaf of Abelian groups on a separable topological space, in which the compact open subsets form a basis for the topology, is injective as a $\mathbf{Z}$-module, if its group of global sections is divisible. Banaschewski refuted this assertion with the help of a counterexample. Essentially the question was already solved by Barthel for sheaves of modules over an arbitrary (fixed) Noetherian ring with unit (cf. G. Barthel, Séminaire Heidelberg–Strasbourg, Institute de rechere mathématique avancée, Strasbourg, 1966/67, Exposé 2).

Theorem 2.8 was proved by the author in 1973 [cf. V. D. Golovin, *Mat. Zametki*, **18**, No. 4, 589–596 (1975)].

16.   Let $\mathcal{O}$ be the sheaf of germs of holomorphic functions in the domain G of the space $\mathbf{C}^n$ and $\mathcal{B}$ be the sheaf of germs of (complex) hyperfunctions on G (cf. Schapira [1]). Then the sheaf $\mathcal{B}$ is an $\mathcal{O}$-module and satisfies the hypotheses of Theorem 2.8 (cf. Schapira

[1, pp. 62 and 107]). Consequently, $\mathscr{B}$ is an injective $\mathcal{O}$-module (cf. Theorem 5.9 of Chapter 5). Further, let $\mathcal{M}$ be the sheaf of terms of meromorphic functions on G (cf. Gunning and Rossi [1, p. 315]). The inclusion $\mathcal{O} \subset \mathscr{B}$ defines a diagram with exact row

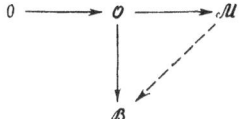

which can be completed by the dashed arrow to a commutative diagram. Here the dashed arrow is a monomorphism, since the extension of $\mathcal{O}$-modules $\mathcal{O} \subset \mathcal{M}$ is essential. Thus, one can consider the sheaf $\mathcal{M}$ as a subsheaf of the sheaf $\mathscr{B}$; in other words, each meromorphic function on G is a hyperfunction.

17.   The results recounted in Section 3 have local character in the following sense: they are proved under the assumption that the coherent analytic sheaves $\mathscr{F}$ and $\mathscr{G}$ considered on the complex manifold X admit free resolutions (1); on the other hand, arbitrary coherent analytic sheaves $\mathscr{F}$ and $\mathscr{G}$ on X satisfy this condition locally, in a neighborhood of each point. This explains the title of the section. The results recounted in it can also be proved under more general assumptions: for example, free resolutions can be replaced by locally free ones (since the present section plays an auxiliary role in the book, we are not striving for maximal generality).

18.   Starting with Serre's papers [1, 2], Lemma 3.3 ("the duality lemma") is invariably used in all papers on duality theory and complex analytic geometry (cf., e.g., Andreotti [2], Andreotti and Banica [1], Andreotti and Kas [2], Banica and Stanasila [4], Ramis and Ruget [1]). The following version of this lemma, establishing a duality between homology and cohomology, is very useful:
        Let E^* be a finite complex in the category of Frechet–Schwartz spaces (with continuous linear differentials). Then the dual spaces $F_n = (E^n)'$, endowed with the strong topology, and the maps dual to the differentials form a chain complex $F_.$ in the category of strong duals to Frechet–Schwartz spaces and there is a natural isomorphism of topological vector spaces

$$\{\tilde{H}^n E^{\cdot}\}' = \tilde{H}_n F_{\cdot},$$

where the dual space is endowed with the strong topology and the tilde denotes factorization by the closure of zero.
        Instead of the category of Frechet–Schwartz spaces one can consider the category of strong duals to Frechet–Schwartz spaces.

Here is a more general form of the duality lemma:

Let $u: E \to F$, $v: F \to G$ be continuous linear maps of separated locally convex spaces and $'u: F' \to E'$, $'v: G' \to F'$ be the dual linear maps. If $v \circ u = 0$ and the space F is semireflexive, then there is a natural isomorphism of vector spaces (without topology)

$$\{\operatorname{Ker} v/\overline{\operatorname{Im} u}\}' = \operatorname{Ker} {}^t u/\overline{\operatorname{Im} {}^t v},$$

where the closure $\overline{\operatorname{Im} u}$ is taken in the original topology of the space F, and the closure $\overline{\operatorname{Im} {}^t v}$ in the strong topology of the space F' (cf. Serre [2, p. 19]).

For the category of locally compact Abelian groups, a duality lemma analogous to Lemma 3.3 was already used in Serre's work on algebraic topology (cf. Aleksandrov [4, p. 16]).

19.   The theory of duality in complex analytic geometry was founded by Serre [2] and Malgrange [3]. The assertions of Corollary 3.5.2 were proved under more general hypotheses than in the text by the author (Golovin [1, 2]; cf. also Banica and Stanasila [4]). The assertion of Corollary 3.5.3 is sometimes called Cartan–Schwartz duality (cf. Serre [1]).

20.   Essentially Lemma 3.6 was proved first by Malgrange [6]. We have borrowed the formulation and proof of Lemma 3.6 from Ramis and Ruget [1]. Andreotti and Banica [1, p. 1008] proved a more general lemma:

Any continuous linear morphism of complexes in the category of Frechet spaces (or of strong duals to Frechet–Schwartz spaces), which is an algebraic quasiepimorphism, is a topological quasiepimorphism.

Another lemma which was essentially proved by Weil [1] is also known, and is sometimes used for the same purposes as Lemma 3.6:

In the commutative diagram of topological vector spaces and continuous linear maps

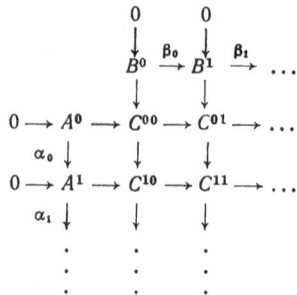

let all rows and columns starting with the second be exact, and the maps which appear in them be homomorphisms. Then the first column and first row define cochain complexes $A^{\cdot}$ and $B^{\cdot}$, respectively, for which, for each $k = 0, 1, \ldots$ the cohomology topological spaces $H^k A^{\cdot}$ and $H^k B^{\cdot}$ are naturally isomorphic, and the coboundary operators α_k and β_k can be homomorphisms only simultaneously [cf. V. D. Golovin, *Ukr. Geom. Sb.*, **13**, 27–63 (1973); a somewhat more general assertion is proved in V. P. Palamodov, *Linear Differential Operators with Constant Coefficients* [in Russian], Nauka, Moscow (1967)].

21. The separation of the topological vector spaces $\text{Ext}_{\mathcal{O}}^k(X; \mathcal{F}; \mathcal{G})$ follows from its finite-dimensionality in view of Lemma 4.5 of Chapter 3 (cf. Serre [2, p. 21]).

22. The first isomorphism of Corollary 3.12.2 was first proved by the author [V. D. Golovin, *Funkts. Anal. Ego Prilozhen.*, **5**, No. 4, 66 (1971); cf. also Banica and Stanasila [2], Malgrange [7], and Golovin, *Teor. Funkts., Funkts. Anal. Ikh Prilozhen.*, **16**, 74–78 (1972)]. The second isomorphism of Corollary 3.12.2 and the first part of Corollary 3.12.3 were first proved by Schapira [2]. The second part of Corollary 3.12.3 was proved even earlier by Martineau [1].

23. Let $\mathcal{O}$ be the sheaf of germs of holomorphic functions in the domain G of the space $\mathbf{C}^n$. It follows in particular from Theorem 4.4 that $\text{inj dim}_{\mathcal{O}} \mathcal{O} = n$. One can give a different, shorter proof of this assertion, based on the theory of hyperfunctions (cf. Golovin [3]). It is known that there exists a resolution

$$0 \longrightarrow \mathcal{O} \longrightarrow \mathcal{B}^{0,\,0} \xrightarrow{d''} \mathcal{B}^{0,\,1} \xrightarrow{d''} \ldots \xrightarrow{d''} \mathcal{B}^{0,\,n} \longrightarrow 0$$

(cf. Schapira [1, p. 73]), where $\mathcal{B}^{0,k}$ is the sheaf of germs of different forms of double degree $(0, k)$, whose coefficients are hyperfunctions. By Theorem 2.8 the $\mathcal{O}$-modules $\mathcal{B}^{0,k}$ are injective. Consequently, $\text{inj dim}_{\mathcal{O}} \mathcal{O} \leq n$. On the other hand, $H_c^n(G; \mathcal{O}) \neq 0$ by Corollary 3.5.3; hence, $\text{inj dim}_{\mathcal{O}} \mathcal{O} \geq n$. The assertion is proved.

24. The assertion of Corollary 4.6 was first proved by Milnor [1] in a somewhat different form. A generalization of the Milnor exact sequence was found by Kh. N. Inasaridze [cf. *Soobshch. Akad. Nauk GSSR*, **79**, No. 1, 17–20 (1975)].

25.   Let $\mathcal{O}$ be the sheaf of germs of holomorphic functions on the domain
        G of the space $\mathbf{C}^n$. Then, as follows from Hilbert's syzygy theorem
        (cf. Gunning and Rossi [1, p. 97]), gl dim $\mathcal{O}_z = n$ for each point $z \in$
        G.

26.   Theorem 4.9 on the global dimension of the sheaf of germs of
        holomorphic functions was first proved by the author [cf. Golovin
        [3]; cf. also Golovin, *Teor. Funkts., Funkts. Anal. Ikh Prilozhen.*,
        32, 22–28 (1979)] with the help of the theory of hyperfunctions
        (Schapira [1]) and Malgrange's theorem on division of distributions
        (Malgrange [5]). In the text there is given a new proof of Theorem
        4.9, not using any results on hyperfunctions and distributions and
        which, in this sense, is elementary.

27.   The groups $H_k^c(\mathfrak{u}; F)$ are nothing but the homology groups of P. S.
        Aleksandrov of the nerve of the covering $\mathfrak{u}$ with coefficients in the
        copresheaf F (cf. Note 3).

28.   Elementary copresheaves and projective resolutions of copre-
        sheaves were first introduced by Deheuvels [1, 2].

29.   Let $F = \mathscr{A}'$ be the copresheaf of analytic functionals on the space
        $\mathbf{R}^n$. In other words, for each open set $U \subset \mathbf{R}^n$

$$F(U) = \{\mathcal{A}(U)\}'$$

        is the vector space of all continuous linear forms on the topological
        vector space $\mathscr{A}(U)$ of all (real) analytic functions on U. Then the
        sheaf $\mathscr{G}$ associated with the copresheaf F coincides with the sheaf $\mathscr{B}$
        of germs of hyperfunctions on $\mathbf{R}^n$. In fact, for each compact set M
        $\subset \mathbf{R}^n$

$$G(M) = \mathcal{A}'(\mathbf{R}^n)/\mathcal{A}'(\mathbf{R}^n \setminus M).$$

        On the other hand, for each relatively compact open set $U \subset \mathbf{R}^n$

$$\mathscr{B}(U) = \mathcal{A}'(\mathbf{R}^n)/\mathcal{A}'(\mathbf{R}^n \setminus U)$$

        (cf. Schapira [1, p. 58). It follows easily from this that $\mathscr{G}_x = \mathscr{B}_x$
        for each point $x \in \mathbf{R}^n$.

30. The basic results of Section 5 were first recounted briefly in a note of the author [cf. Golovin [5]; for a detailed account, see Golovin, *Teor. Funkts., Funkts. Anal. Ikh Prilozhen.*, **33**, 26–36 (1980)].

Chapter 2

1. The definition of the sheaf of germs of homology can be formulated briefly as follows (cf. Golovin [4]). Let X be an arbitrary complex space and $\mathscr{F}$ be an arbitrary analytic sheaf on X. Then by Theorem 1.2, for each integer k there exists an analytic sheaf $\mathscr{H}_k(\mathscr{F})$ on X, which is unique up to isomorphism, which has the following property: if M is an open set in X, realized as an analytic set in the domain G of the space $\mathbf{C}^n$, then on M there is a natural isomorphism

$$\mathscr{H}_k(\mathscr{F}) = \mathscr{E}\mathrm{xt}^{n-k}_{\mathcal{O}_G}(\mathscr{F}^G, \Omega^n_G),$$

where $\mathcal{O}_G$ is the sheaf of germs of holomorphic functions on G, Ω_G^n is the sheaf of germs of holomorphic n-forms on G, $\mathscr{F}^G$ is the trivial extension of the sheaf $\mathscr{F}$ to the domain G. The sheaf $\mathscr{H}_k(\mathscr{F})$ is called the *sheaf of germs of homology* of dimension k of the analytic sheaf $\mathscr{F}$.

2. One should note that for a coherent analytic sheaf $\mathscr{F}$ on the complex space X the existence of the sheaf of germs of homology $\mathscr{H}_k(\mathscr{F})$ can be proved considerably more briefly (cf. point 1.2.2). Essentially this was done by Andreotti and Kas [1]. In fact, they defined for each integer $k \geq 0$ the *dualizing sheaf* $\mathscr{D}^k\mathscr{F}$ on X, generated by the presheaf $U \longmapsto \{H_c^k(U; \mathscr{F})\}'$. If the open set $M \subset X$ is realized as an analytic set in the domain G of the space $\mathbf{C}^n$, then by virtue of Serre duality on M there is a natural isomorphism of analytic sheaves

$$\mathscr{D}^k\mathscr{F} = \mathscr{E}\mathrm{xt}^{n-k}_{\mathcal{O}_G}(\mathscr{F}^G, \Omega^n_G).$$

$$(*)$$

Essentially, this fact was known even before the work of Andreotti and Kas [1] (cf. Banica and Stanasila [1], Golovin [2]). Andreotti and Kas used the isomorphism (*) to prove the coherence of the sheaf $\mathscr{D}^k\mathscr{F}$: since the right side of (*) is a coherent analytic sheaf on G, the left side, i.e., $\mathscr{D}^k\mathscr{F}$, is a coherent analytic sheaf on M. However, to prove the existence of the sheaf $\mathcal{H}^k(\mathcal{F})$ it is necessary to use the implication in the opposite direction: since the left side in (*) is independent of how the analytic set M is realized in the domain G of the space $\mathbf{C}^n$, the right side also is independent of this. In particular, there is an isomorphism of analytic sheaves $\mathcal{D}^k\mathcal{F} = \mathcal{H}^k(\mathcal{F})$ on X. Unfortunately, in general, i.e., for an arbitrary analytic sheaf $\mathcal{F}$ on X, these arguments are inapplicable, and to prove the existence of the sheaves of germs of homology $\mathcal{H}^k(\mathcal{F})$ it is necessary to use Theorem 1.2 in its full generality.

3. Sheaves of germs of homology and homology groups of analytic sheaves were first introduced in a paper by the author (cf. Golovin [4]). Here, too, one should note that without the results of local character previously found by the author and recounted in Chapter 1 (injectivity criterion, local duality, injective dimension, homological properties of fine sheaves), a satisfactory definition of the homology groups of analytic sheaves would be impossible.

4. Thus, the sheaf of germs of homology $\mathcal{H}^k(\mathcal{F})$ is generated by the presheaf of homology groups $U \mapsto H_k(U; \mathcal{F})$. This justifies the appelation of the sheaf $\mathcal{H}^k(\mathcal{F})$.

5. It follows from Theorem 2.4 that the definition of the homology groups of a complex space X with coefficients in the analytic sheaf $\mathcal{F}$ can be represented in the following form:

$$H_k^\Phi(X; \mathcal{F}) = H_k H_0^\Phi(X; \mathscr{C}^\bullet(\mathcal{F})).$$

This definition is completely analogous to the (dual) definition of co-homology groups a la Godement (cf. point 1.7.3 of Chapter 1):

$$H_\Phi^k(X; \mathcal{F}) = H^k H_\Phi^0(X; \mathscr{C}^\bullet(\mathcal{F}))$$

(now here X is an arbitrary topological space, $\mathcal{F}$ is an arbitrary sheaf of Abelian groups on X).

6. Theorem 4.1 is an analog of a familiar theorem of Cartan [2], which connects the singular homology of a topological space with its Aleksandrov–Čech cohomology.

7. Corollary 4.2 is an analog of the classical Poincaré duality in algebraic topology [cf. H. Poincaré, "Complément à l'analysis situs," *Rend. Circ. Mat. Palermo*, **13**, 285–343 (1899); Aleksandrov [5]; Cartan [2]; A. Borel, "The Poincaré duality in generalized manifolds,"

Michigan Math. J., **4**, No. 3, 227–239 (1957); G. E. Bredon, *Sheaf Theory*, McGraw-Hill, New York (1967); E. G. Sklyarenko, *Izv. Akad. Nauk SSSR, Ser. Mat.*, **35**, No. 4, 831–843 (1971)].

8. Proposition 4.3 and its proof were communicated to the author by E. G. Sklyarenko.

9. Dualizing complexes are the basic apparatus in the papers of Ramis and Ruget on duality theory and its applications (cf. Ramis and Ruget [1-3]; Ramis, Ruget, and Verdier [1]; Ramis [1-3]). Dualizing complexes are not used anywhere in the present book. Hence, one can omit Section 5 on a first reading.

10. In Ramis and Ruget [1], a number of requirements are imposed on a dualizing complex, among which the most important (along with conditions a) and b)) are the following:
    c) if G is a domain in the space $\mathbf{C}^n$, then the complex $\mathcal{K}_G^\cdot$ is a resolution of the complex $\Omega_{G}^n[n]$;
    d) if the complex space X is finite-dimensional, then the complex $\mathcal{K}_X^\cdot$ is bounded, namely: $\mathcal{K}_X^p = 0$ for $p < -\dim_{\mathbf{C}} X$ and for $p > 0$.
    The remaining conditions of Ramis and Ruget are easily derived from conditions a)-d).

11. Dualizing complexes first appeared in algebraic geometry, namely: in connection with duality theory for the cohomology of coherent algebraic sheaves Grothendieck [3] constructed the so-called *Cousin complex* $\mathcal{K}_X^\cdot$ on a scheme X of finite type over an arbitrary ground field k (cf., e.g., Shafarevich [1]), which he called the complex of residues of the scheme X or the complex of "generalized local Cousin elements," and which consists of algebraic quasicoherent sheaves and is the injective dualizing complex of the scheme X. General principles of construction of Cousin complexes with the help of local cohomology on schemes were developed by Hartshorne [1], and on arbitrary topological spaces by Suominen [2].
    Ramis and Ruget [1] transferred the apparatus of dualizing complexes to analytic geometry. They first constructed a dualizing complex $K_X^\cdot$ on a complex space X of finite dimension, which is countable at infinity. The construction of the complex $K_X^\cdot$ closely follows the idea of Grothendieck and in broad strokes reduces to the following. If G is a domain in the space $\mathbf{C}^n$, then for each point $z \in G$ there is defined the stalk $K_{G,z}^\cdot$ of the Cousin complex (in the sense of Grothendieck) on the scheme Spec $\mathcal{O}_{G,z}$ (cf., e.g., Shafarevich [1]), and afterwards all the stalks $K_{G,z}^\cdot (z \in G)$ are tied together into a

sheaf $K_G^{\cdot}$. If M is an analytic set in the domain G of the space $\mathbf{C}^n$, then the complex $K_M^{\cdot}$ is defined naturally with the help of condition b) of point 5.1. As a result one gets a dualizing complex $K_X^{\cdot}$, satisfying conditions a)-d).

Fouché [1, 2] constructed another dualizing complex $\hat{K}_X^{\cdot}$ on a complex space X of finite dimension which is countable at infinity (cf. also Banica and Stanasila [4, p. 93]). If G is a domain in the space $\mathbf{C}^n$, then the complex $\hat{K}_G^{\cdot}$ is defined in the familiar way as the Cousin complex on G (cf. Suominen [2]). If M is an analytic set in the domain G of the space $\mathbf{C}^n$, then the complex $\hat{K}_M^{\cdot}$ is defined in the usual way with the help of condition b) of point 5.1. The dualizing complex $\hat{K}_X^{\cdot}$ constructed in this way also satisfies conditions a)-d).

There exists a natural injective morphism of complexes of $\mathcal{O}_X$-modules

$$K_X^{\cdot} \longrightarrow \hat{K}_X^{\cdot},$$

which is a quasiisomorphism (cf. Fouché [1, 2]). On the other hand, if G is a domain in the space $\mathbf{C}^n$, then the sheaf $\hat{K}_G^{-n}$ is generated by the presheaf

$$U \longmapsto \varinjlim \Gamma\,(U \diagdown S;\; \Omega_G^n),$$

where the inductive limit is taken with respect to the filtered set, ordered by inclusion, of all analytic subsets $S \subset U$ of dimension $n - 1$. It follows directly from this that the dualizing complex $\hat{K}_G^{\cdot}$ (and hence also $K_G^{\cdot}$) is not soft, hence not injective. Stanasila [1] notes as a useful property of the dualizing complex $\hat{K}_X^{\cdot}$ the existence of a natural topology on the group of its sections. However, one can show that this natural topology is trivial [cf. V. D. Golovin, *Mat. Sb.*, **73**, No. 1, 21–41 (1967)].

Ruget [1] proved (with the help of the theory of residues in the sense of Herrera) that on a complex manifold X of dimension n there exists a natural injective morphism of complexes of $\mathcal{O}_X$-modules

$$K_X^{-p} \longrightarrow {}'\mathcal{D}_X^{n,\,n-p},$$

which is a quasiisomorphism (cf. also Ramis and Ruget [3]), where ${}'\mathcal{D}_X^{n,\cdot}$ is the Dolbeault complex defined by distributions (i.e., consisting of currents) on X. Fouché [2] also proved that on a complex manifold X of dimension n there exists a natural injective morphism of complexes of $\mathcal{O}_X$-modules

$$\hat{K}_X^{-p} \longrightarrow \mathscr{B}_X^{n,\ n-p},$$

which is a quasiisomorphism [cf. Z. Mebkhout, "Local cohomology of analytic spaces, " *Publ. R. I. M. S. Kyoto Univ.*, **12**, Suppl., 247–256 (1977)], where $\mathscr{B}_X^{n,\cdot}$ is the Dolbeault complex defined by hyperfunctions (i.e., consisting of hypercurrents) on X. Ramis [4] interprets the dualizing complexes $K_X^\cdot$ and $\hat{K}_X^\cdot$ on a complex manifold X as Dolbeault complexes defined, respectively, by algebraic distributions and algebraic hyperfunctions on X.

12. Let $\mathcal{O}$ be a sheaf of commutative rings with unit on the topological space X and $\mathscr{K}^\cdot$ be an arbitrary complex of $\mathcal{O}$-modules. A complex of $\mathcal{O}$-modules $\mathscr{L}^\cdot$ is called a *resolution* of the complex $\mathscr{K}^\cdot$ if there exists a morphism of complexes $\mathscr{K}^\cdot \to \mathscr{L}^\cdot$, which is a quasiisomorphism, i.e., which induces, for each p, an isomorphism $\mathcal{O}$-modules of cohomology $H^p\mathscr{K}^\cdot = H^p\mathscr{L}^\cdot$. The complex $\mathscr{L}^\cdot$ is called *injective* if all the $\mathcal{O}$-modules $\mathscr{L}^p$ are injective. Let $\mathscr{L}^\cdot$ be an injective resolution of the complex $\mathscr{K}^\cdot$. For an arbitrary additive covariant left exact functor F on the category of $\mathcal{O}$-modules we set

$$R^p F \left(\mathscr{K}^\cdot \right) = H^p F \left(\mathscr{L}^\cdot \right)$$

(cf. Hartshorne [1], Verdier [2]). Up to isomorphism this definition is independent of the choice of injective resolution $\mathscr{L}^\cdot$.

13. The basic goal of this section (i.e., Section 5) is the investigation of the connection between dualizing complexes and the homology of analytic sheaves. Theorem 5.4 and its corollary show how one can calculate the homology of analytic sheaves with the help of dualizing complexes, and Theorem 5.5 gives a general method for constructing (injective) dualizing complexes with the help of the homology of analytic sheaves.

14. The dualizing complex constructed by Grothendieck [3] is injective, and the dualizing complexes constructed by Ramis and Ruget [1] and Fouché [1] are not injective. Theorem 5.5 shows that on complex spaces there exist injective dualizing complexes. Theorem 5.5 gives a general method for constructing injective dualizing complexes: it suffices to give on each complex space X a locally fine resolution $\mathscr{L}_X^\cdot$ of the structure sheaf $\mathcal{O}_X$, consisting of flat $\mathcal{O}_X$-modules, and satisfying the natural compatibility conditions. As such a resolution $\mathscr{L}_X^\cdot$ one can

take the canonical resolution of Godement $\mathscr{C}^{\cdot}(\mathcal{O}_X)$, but there are also other possibilities.

For example, we associate with each open set $U \subset X$ the group of maps $U^{k+1} \to \Gamma(U; \mathcal{O}_X)$. Defining the restriction map in the natural way, we get a presheaf on X. The sheaf $\mathscr{F}^k(\mathcal{O}_X)$ generated by this presheaf is called *the sheaf of terms of Alexander–Spanier cochains* of degree k on X, with values in $\mathcal{O}_X$. In the natural way one defines homomorphisms of sheaves d: $\mathscr{F}^k(\mathcal{O}_X) \to \mathscr{F}^{k+1}(\mathcal{O}_X)$ and an inclusion $\mathcal{O}_X \to \mathscr{F}^0(\mathcal{O}_X)$. We get a resolution

$$0 \longrightarrow \mathcal{O}_X \longrightarrow \mathscr{F}^0(\mathcal{O}_X) \xrightarrow{d} \mathscr{F}^1(\mathcal{O}_X) \xrightarrow{d} \ldots$$

(cf. Godement [1, p. 156]), in which the sheaves $\mathscr{F}^k(\mathcal{O}_X)$ are locally fine flat $\mathcal{O}_X$-modules (cf. Godement [1, p. 181]), i.e., this resolution satisfies the hypotheses of Theorem 5.5.

We associate with each open set $U \subset X$ the group of maps $S_k(U) \to \Gamma(U; \mathcal{O}_X)$, where $S_k(U)$ is the group of k-dimensional singular chains of the space U. Defining the restriction maps in the obvious way, we get a presheaf on X. The sheaf $\mathscr{S}^k(\mathcal{O}_X)$ generated by this presheaf is called *the sheaf of germs of localized singular k-cochains* of the space X with values in $\mathcal{O}_X$ (cf. Godement [1, p. 184]). The sheaves $\mathscr{S}^k(\mathcal{O}_X)$ are locally fine flat $\mathcal{O}_X$-modules (cf. Godement [1, p. 185]). Thus, with the help of the coboundary operator δ we get another resolution of the structure sheaf $\mathcal{O}_X$

$$0 \longrightarrow \mathcal{O}_X \longrightarrow \mathscr{S}^0(\mathcal{O}_X) \xrightarrow{\delta} \mathscr{S}^1(\mathcal{O}_X) \xrightarrow{\delta} \ldots,$$

satisfying the hypotheses of Theorem 5.5.

One can also define various "Cartan resolutions" (cf. Remark 3.1), which will satisfy the hypotheses of Theorem 5.5.

Chapter 3

1. Let X be a complex space which is countable at infinity, Φ be an admissible family of supports in X, $\mathfrak{u}$ be a locally infinite open covering of the space X, adapted to Φ. Then for any analytic sheaf $\mathscr{F}$ on X there exists a spectral sequence

$$E^2_{p,q} = H^{\Phi}_p(\mathfrak{u}; \mathscr{H}^c_q(\mathscr{F})) \Rightarrow H^{\Phi}_{p+q}(X; \mathscr{F}),$$

where $\mathcal{H}_q{}^c(\mathcal{F})$ is the copresheaf $U \mapsto H_q{}^c(U; \mathcal{F})$ on the category of open sets in X [cf. V. D. Golovin, *Usp. Mat. Nauk*, **36**, No. 1, 59–71 (1981)].

2. It is easy to see that the two topologies on the vector space $\Gamma(U_{i_0 \ldots i_k}; \mathcal{F})$, defined, respectively, in the book of Gunning and Rossi [1, p. 303] and in point 3.1 of Chapter 1, actually coincide.

3. As will follow from Theorem 4.1 (or from Theorem 2.3), the assertion of Lemma 2.2.1 is also valid without the assumption that the covering $\mathfrak{u}$ is sufficiently fine.

4. Theorem 2.3 was first proved by the author [cf. V. D. Golovin, *Teor. Funkts., Funkts. Anal. Ikh. Prilozhen.*, **25**, 49–56 (1976)].

5. Serre [2] first proved the duality theorem for a locally free sheaf $\mathcal{F}$ on a complex manifold X of dimension n, countable at infinity, in the following form: if the topological vector spaces $H^k(X; \mathcal{F})$ and $H^{k+1}(X; \mathcal{F})$ are separated, then there exists a natural isomorphism of vector spaces (without topology)

$$\{H^k(X; \mathcal{F})\}' = H_c^{n-k}\left(X; \mathcal{H}om_{\mathcal{O}}(\mathcal{F}, \Omega_X^n)\right)$$

(the fact that this is an isomorphism of topological vector spaces was first proved by Laufer [1]). This duality is defined by the natural pairing, which has an intuitive cohomological interpretation (cf. Serre [2]). The special case of the Serre duality theorem for a free sheaf $\mathcal{F}$ on a holomorphically complete manifold X was known somewhat earlier (Serre [1] attributes this result to Cartan and Schwartz). Simultaneously Serre proved an analogous duality theorem in algebraic geometry [cf. O. Zariski, "Algebraic sheaf theory," *Bull. Am. Math. Soc.*, **62**, 117–141 (1956)]. Grothendieck's papers [1, 3] completely determined the utmost development of these investigations in algebraic geometry and turned out to be an essential influence on the corresponding investigations in analytic geometry (i.e., in the theory of complex spaces); in particular, he introduced the functors Ext for sheaves of modules (cf. Grothendieck [1, 2]), dualizing complexes (Cousin complexes), homology of algebraic sheaves (cf. Grothendieck [3]). In analytic geometry Malgrange [3] proved the duality theorem for an arbitrary coherent analytic sheaf $\mathcal{F}$ on a complex manifold X of dimension n, countable at infinity; in the most general form it says: there exists a natural isomorphism of topological vector spaces

$$\{\tilde{H}^k(X;\mathcal{F})\}' = \widetilde{\mathrm{Ext}}^{n-k}_{\mathcal{O}_X,\,c}(X;\mathcal{F},\Omega^n_X)$$

(cf. Golovin [1]), where the tilde denotes factorization by the closure of zero. Malgrange [3] proved this isomorphism only for holomorphically complete or compact manifolds, and his proof was based essentially here on his results on division of distributions (cf. Malgrange [5]); Suominen [1] got these same results again by an elementary method. Obviously Serre's isomorphism is a special case of the Malgrange isomorphism; on the other hand, the Malgrange theorem follows from Theorem 2.3 in view of Proposition 3.4 of Chapter 2. Ramis and Ruget [1] proved two duality theorems on complex spaces. The first of these two theorems says that for an arbitrary coherent analytic sheaf $\mathcal{F}$ on a complex space X of finite dimension, which is countable at infinity, there exists a natural isomorphism of topological vector spaces

$$\{\tilde{H}^k(X;\mathcal{F})\}' = \widetilde{\mathrm{Ext}}^{-k}_{\mathcal{O}_X,\,c}(X;\mathcal{F},K^{\cdot}_X).$$

Thus, essentially one gets a Malgrange isomorphism in which the sheaf $\Omega_X{}^n$ is replaced by the dualizing complex $K_X^{\cdot}$ (constructed by Ramis and Ruget [1]). Ramis and Ruget justify the legality of this substitution as follows: if X is a complex manifold of dimension n, then $K_X^{\cdot}$ is a resolution of the complex $\Omega_X{}^n[n]$; consequently, the complexes $K_X^{\cdot}$ and $\Omega_X{}^n[n]$ are isomorphic in the derived category $D(X)$ of the Abelian category of $\mathcal{O}_X$-modules. Ramis and Ruget [1] also consider the natural pairing, which defines their duality and which has a rather intuitive cohomological interpretation. The "first duality theorem" of Ramis and Ruget follows from Theorem 2.3 in view of Theorem 5.4 of Chapter 2 in strengthened form: first, the complex X is not assumed to be finite-dimensional; second, if $\mathcal{K}_X^{\cdot}$ is an injective dualizing complex, then by the corollary to Theorem 5.4 there exist natural isomorphisms

$$H^c_k(X;\mathcal{F}) = H^{-k}\mathrm{Hom}_{\mathcal{O}_X,\,c}(X;\mathcal{F},\mathcal{K}^{\cdot}_X),$$

i.e., the functors hyper-Ext in the Ramis–Ruget isomorphism can be replaced by essentially simpler invariants. In algebraic geometry duality theorems in precisely this form were proved by Grothendieck [3] nearly thirty years ago (Grothendieck [3] took an isomorphism analogous to that indicated above as the definition of his homology groups of algebraic sheaves).

6. Let X be a complex manifold of dimension n, countable at infinity, and $\mathscr{F}$ be a coherent analytic sheaf on X. Then there is the Malgrange isomorphism of topological vector spaces:

$$\{\tilde{H}_c^k \ (X; \ \mathscr{F})\}' = \widetilde{\text{Ext}}_{\mathcal{O}_X}^{\,n-k} \ (X; \ \mathscr{F}, \ \Omega_X^n)$$

(cf. Golovin [2], V. D. Golovin, *Ukr. Geom. Sb.*, **13**, 27–63 (1973); Banica and Stanasila [4, p. 318]). This isomorphism follows from Theorem 3.3 in view of Proposition 3.4 of Chapter 2. The "second duality theorem" of Ramis–Ruget says that for a coherent analytic sheaf $\mathscr{F}$ on a complex space X of finite dimension, countable at infinity, there is a natural isomorphism of topological vector spaces

$$\{\tilde{H}_c^k(X; \ \mathscr{F})\}' = \widetilde{\text{Ext}}_{\mathcal{O}_X}^{\,-k} \ (X; \ \mathscr{F}, \ K_X^{\cdot})$$

(cf. Ramis and Ruget [1]). This duality theorem can be obtained in strengthened form from Theorem 3.3 in view of Theorem 5.4 of Chapter 2 (cf. Note 5).

7. Andreotti and Kas [2] suggested a new approach to the theory of duality on complex spaces, which apparently originates under the influence of Suominen [1]. The essence of their approach is the following. First of all one introduced the Aleksandrov–Čech homology groups $H_k^{\Psi}(X; F)$ with coefficients in a copresheaf F and with family of supports Ψ. Afterwards, for an arbitrary coherent analytic sheaf $\mathscr{F}$ on the complex space X, countable at infinity, one proves a duality theorem, which has the classical form of the duality between homology and cohomology (cf. Aleksandrov [4, p. 16]) and can be described in the form of a natural isomorphism of vector spaces

$$\{\tilde{H}_{\Phi}^k \ (X; \ \mathscr{F})\}' = \tilde{H}_k^{\Psi} \ (X; \ \mathscr{F}_*), \qquad\qquad (*)$$

where Φ and Ψ are dual families of supports in X, $\mathscr{F}_*$ being the dual copresheaf $U \longmapsto \{\Gamma(U; \ \mathscr{F} \ )\}'$ on the category of open sets in X, where the tilde denotes factorization by the closure of zero (cf. Andreotti and Kas [2, p. 214], and also Andreotti and Banica [1, p. 1033]). Finally, by purely algebraic means (i.e., without the use of the theory of topological vector spaces) one expresses the homology groups $H_k^{\Psi}(X; \ \mathscr{F}_*)$ in terms of familiar functors of the sheaf $\mathscr{F}$. Thus, if X is a complex manifold of dimension n, then one gets a natural isomorphism of vector spaces

$$H_k^{\Psi} \ (X; \ \mathscr{F}_*) = \text{Ext}_{\mathcal{O}_X, \ \Psi}^{\,n-k} \ (X; \ \mathscr{F}, \ \Omega_X^n)$$

(cf. Andreotti and Kas [2, p. 222]) and the isomorphism (*) becomes the Malgrange duality theorem; if X is a complex space of finite dimension, then one gets a natural isomorphism of vector spaces

$$H_k^{\Psi}(X;\ \mathscr{F}_*) = \mathbf{Ext}_{\mathfrak{G}_X,\ \Psi}^{-k}(X;\ \mathscr{F},\ K_X^{\cdot}),$$

where $K_X^{\cdot}$ is a dualizing complex of the space X (cf. Andreotti and Kas [2, p. 261]), and the isomorphism (*) becomes the Ramis–Ruget duality theorem; under the same hypotheses, with the help of a dualizing complex one gets a spectral sequence

$$E_{p,\ q}^2 = H_{\Psi}^{-p}(X;\ \mathscr{D}^q\mathscr{F}) \Rightarrow H_{p+q}^{\Psi}(X;\ \mathscr{F}_*), \qquad (**)$$

where $\mathscr{D}^q\mathscr{F}$ is a dualizing sheaf (cf. Andreotti and Kas [2, p. 259], Andreotti and Banica [1, p. 1040]; cf. Note 2 to Chapter 2 for the definition of the sheaves $\mathscr{D}^q\mathscr{F}$), so from the isomorphism (*) one gets a generalization of the duality theorem of Andreotti and Kas [1]. If U is a holomorphically complete open set in the complex space X, then there is a natural isomorphism of vector spaces

$$\{\Gamma(U;\ \mathscr{F})\}' = H_0^c(U;\ \mathscr{F})$$

(cf. Theorem 2.3 of Chapter 3); in other words, for the dual copresheaf $\mathscr{F}_*$ of Andreotti–Kas on the category of holomorphically complete open sets in X there is a natural isomorphism $\mathscr{F}_* = \mathscr{H}_0^c(\mathscr{F})$. Consequently, if $\mathfrak{u}$ is a locally finite covering of the space X by holomorphically complete open sets, adapted to the admissible family of supports Ψ, then there exists a natural isomorphism of vector spaces

$$H_k^{\Psi}(X;\ \mathscr{F}) = H_k^{\Psi}(\mathfrak{u};\ \mathscr{F}_*).$$

Thus, from Theorems 2.3 and 3.3 of Chapter 3, we get the isomorphism (*) for the families of supports consisting, respectively, of all closed and all compact sets in X; analogously one can get the isomorphism (*) in general [cf. V. D. Golovin, *Usp. Mat. Nauk*, **36**, No. 1, 59–71 (1981)]. As to the spectral sequence (**), it is now obtained directly from Theorem 4.1 of Chapter 2; here it is not necessary to assume that the space X is finite-dimensional. Thus we see that the homology of analytic sheaves, the theory of which is developed in turn in this book, is a natural instrument in the theory of duality on complex spaces, more flexible than the functors hyper-Ext, used by Ramis–Ruget, or the homology of Aleksandrov–Čech with coefficients in copresheaves, used by Andreotti and Kas.

8. If X is a complex manifold of dimension n, then from Theorems 4.1 and 4.6, in view of Proposition 3.4 of Chapter 2, one gets the Malgrange duality theorems [3] (cf. also Suominen [1]), and from Theorem 4.4, the duality theorem of Banica and Stanasila [1].

9. The assertion of Theorem 4.4 becomes false generally if one replaces the family of all compact subsets of X by an arbitrary paracompactifying family of supports. We consider, for example, the domain of holomorphy in C^2

$$X = \{(z_1, z_2) \in C^2 : z_1 z_2 \neq 0\}.$$

Then the sets

$$S_r = \{z_1 \neq 0; \ 0 < |z_2| \leqslant r\} \cup \{0 < |z_1| \leqslant r; \ z_2 \neq 0\}$$

for $r > 0$ and all their closed subsets form an admissible family of supports Φ in X. However, the topological vector space $H_\Phi^1(X; \mathcal{O}_X) \neq 0$ is not separable [cf. V. D. Golovin, *Usp. Mat. Nauk*, **36**, No. 1, 59–71 (1981)].

10. Lemma 4.5 and its proof are due to Serre [2].

11. Andreotti and Kas [1] suggested the following generalization of Serre–Malgrange duality. Let $\mathcal{F}$ be a coherent analytic sheaf on the compact complex space X. Then there is a spectral sequence

$$E^2_{p, q} = H^{-p}(X; \mathcal{D}^q \mathcal{F}) \Longrightarrow \{H^{p+q}(X; \mathcal{F})\}',$$

where $\mathcal{D}^q \mathcal{F}$ are dualizing sheaves (cf. Note 2 to Chapter 2 for their definition). In view of Theorem 4.1 of Chapter 2, this assertion follows directly from Theorem 4.6. See Note 7 to Chapter 3 for a more general assertion.

12. Theorem 4.6 recalls the duality theorem in algebraic topology for the homology and cohomology vector spaces of finite simplicial complexes with coefficients in a field (cf. Aleksandrov [4, p. 16], and also Lefschetz [2, p. 167]).

13. A more general assertion is formulated in Note 1 to Chapter 3 [cf. V. D. Golovin, *Usp. Mat. Nauk*, **36**, No. 1, 59–71 (1981)].

14. The spectral sequence in Theorem 5.5 is nothing but the Leray spectral sequence of a proper morphism of complex spaces for the homology

groups of analytic sheaves. Theorem 5.5 was first proved by the author [cf. Golovin [4], V. D. Golovin, *Izv. Vyssh. Uchebn. Zaved., Mat.*, No. 7, 15–21 (1979)]. If the space Y consists of one point (in this case the space X is compact) and is reduced, then the spectral sequence degenerates. One gets a natural isomorphism of vector spaces

$$\mathrm{Hom}_C(H^k(X; \mathscr{F}), \ C) = H_k(X; \ \mathscr{F}),$$

which is equivalent to the assertion of Theorem 4.6. For arbitrary morphisms of complex spaces Theorem 5.5 is false. In fact, let $X = \mathbf{C}^n$, Y be a one-point reduced space, f be a constant map. If Theorem 5.5 were true in this case, we would get an isomorphism

$$\mathrm{Hom}_C(\Gamma(X; \mathcal{O}_X), \ C) = H_0(X; \mathcal{O}_X).$$

On the other hand, by Poincaré duality (cf. Corollary 4.2 of Chapter 2) we have an isomorphism $H_0(X; \mathcal{O}_X) = H^n(X; \Omega_X^n)$, while by Cartan's Theorem (B) the last group is trivial. We have found a contradiction.

15. Corollary 5.6 essentially establishes the compatibility of homology with direct images. Ramis, Ruget, and Verdier [1] proved that for a proper morphism of complex spaces $f\colon X \to Y$, where Y is an open polydisc in the space $\mathbf{C}^n$, and an arbitrary coherent analytic sheaf $\mathscr{F}$ on X, there exists a natural isomorphism

$$Rf_* R\,\mathscr{H}\mathrm{om}_{\mathcal{O}_X}(\mathscr{F}, \ K_X^{\cdot}) = R\mathscr{H}\mathrm{om}_{\mathcal{O}_Y}(Rf_*\mathscr{F}, \ K_Y^{\cdot})$$

in an arbitrary category (cf. also Ramis and Ruget [3]). This is a relative duality theorem, establishing the compatibility of the dualizing functor with the direct image functor (cf. Hartshorne [1, p. 379]); essentially it is equivalent to Theorem 5.5 or Corollary 5.6 (which, however, are proved for an arbitrary complex space Y). Ramis and Ruget [3] proved a more general relative duality theorem:

Let $f\colon X \to Y$ be an arbitrary morphism of complex spaces, where X is a complex space, countable at infinity, with bounded Zariski dimension, and Y is an open polydisc in the space $\mathbf{C}^n$. Then for any coherent analytic sheaf $\mathscr{F}$ on X, there exist:

a) bounded complexes $\mathscr{M}^{\cdot}$ and $\mathscr{N}^{\cdot}$, respectively *FN*-free and *DFN*-free $\mathcal{O}_Y$-modules with continuous differentials, representing, respectively, $Rf_*\mathscr{F}$ and $Rf_!R\mathscr{H}\mathrm{om}_{\mathcal{O}_X}\mathscr{F}, K_X^{\cdot})$ in the derived category;

b) a continuous quasiisomorphism

$$\mathcal{N}^\bullet \longrightarrow \mathcal{H}omtop_{\mathcal{O}_Y}\left(\mathcal{M}^\bullet,\ \Omega^n_Y\,[n]\right),$$

representing the natural isomorphism

$$Rf_!\,R\mathcal{H}om_{\mathcal{O}_X}\left(\mathcal{F},\ K^\bullet_X\right)=R\mathcal{H}omtop_{\mathcal{O}_Y}\left(Rf_*\mathcal{F},\ K^\bullet_Y\right)$$

in the derived category (cf. Ramis and Ruget [3. p. 117]).

There is also an analogous theorem in which $f_!$ and f_* exchange places (cf. Ramis and Ruget [3, p. 129]).

Chapter 4

1. By the famous theorem of P. S. Uryson, the space X is metrizable [cf., e.g., P. S. Aleksandrov, *Introduction to Set Theory and General Topology* [in Russian], Nauka, Moscow (1977)].

2. This construction, the idea of which goes back as far as Aleksandrov [1], was studied in detail by Sklyarenko [1] in connection with his approach to the Steenrod–Sitnikov homology theory.

3. Theorem 3.3 was first proved by the author in 1974 (cf. Golovin [6]). If X is a complex manifold of dimension n which is countable at infinity, S is a closed set in X, and $\mathcal{F}$ is a coherent analytic sheaf on X, then there is a natural isomorphism of topological vector spaces

$$\{\tilde{H}^k_S(X;\ \mathcal{F})\}'=\widetilde{\mathrm{Ext}}^{n-k}_{\mathcal{O}_X,\ c}\left(S;\ \mathcal{F},\ \Omega^n_X\right)$$

    (cf. V. D. Golovin, *Funkts. Anal. Ego Prilozhen.*, **5**, No. 4, 66; *Teor. Funkts., Funkts. Anal. Ikh Prilozhen.*, **16**, 74–78 (1972); this isomorphism follows from Theorem 3.3 by virtue of Proposition 3.4 of Chapter 2.

4. If X is a complex manifold, then in view of Proposition 3.4 of Chapter 2 from Theorem 3.5 one gets the duality theorem of Banica and Stanasila [2] (cf. also Malgrange [7]).

5. The assertion of the corollary to Theorem 3.5 was first proved by Banica [1] by a different, less direct method (cf. also Andreotti and Banica [1, p. 1169]).

6. Theorem 3.8 was first proved by the author in 1974 (cf. Golovin [6]). If X is a complex manifold of dimension n which is countable at infinity, S is a closed set in X, and $\mathcal{F}$ is a coherent analytic sheaf on X, then there is a natural isomorphism of topological vector spaces

$$\{\tilde{H}_c^k (S;\ \mathcal{F})\}' = \widetilde{\mathrm{Ext}}_{\mathcal{O}_X,\ S}^{n-k} (X;\ \mathcal{F},\ \Omega_X^n)$$

(cf. Schapira [2]; for a special case, cf. Martineau [1]). This isomorphism follows from Theorem 3.8 in view of Proposition 3.4 of Chapter 2.

Duality theorems respectively for local cohomology spaces and cohomology spaces of a closed set with coefficients in coherent analytic sheaves, different in form from Theorems 3.3 and 3.8, were proved by Andreotti and Banica [1]. Their paper is a continuation and natural development of the paper of Andreotti and Kas [2], and like the latter, is based on the systematic use of (relative) Aleksandrov–Čech homology groups with coefficients in copresheaves (cf. Note 7 to Chapter 3 in this connection).

7. *Alexander–Pontryagin Duality.* Let X be a complex space which is countable at infinity, S be a closed set in X, $\mathcal{F}$ be a coherent analytic sheaf on X. Then for each integer k there are defined natural continuous linear maps

$$\alpha:\quad H^k (X;\ \mathcal{F}) \longrightarrow H^k (X \setminus S;\ \mathcal{F}),$$
$$\beta:\ H_{k+1}^c (X;\ \mathcal{F}) \longrightarrow H_{k+1}^c (S;\ \mathcal{F}).$$

The following theorem holds.

In order that the topological vector space

$$H_{k+1}^c (S;\ \mathcal{F}) / \beta H_{k+1}^c (X;\ \mathcal{F})$$

be naturally isomorphic to the strong dual to the topological vector space

$$H^k (X \setminus S;\ \mathcal{F}) / \overline{\alpha H^k (X;\ \mathcal{F})},$$

it is necessary and sufficient that the continuous linear map

$$H_S^{k+1} (X;\ \mathcal{F}) \longrightarrow H^{k+1} (X;\ \mathcal{F})$$

have closed kernel.

The proof of the theorem is based on Theorem 3.3 [cf. V. D. Golovin, *Mat. Zametki*, **13**, No. 4, 561–564 (1973)]. The condition of the theorem automatically holds if the topological vector space $H^{k+1}(X;\ \mathcal{F})$ is separated.

As a direct consequence of the theorem we get:

If X is a complex manifold of dimension n and $\mathcal{F}$ is a locally free analytic sheaf on X, then the topological vector space

$$H_c^{n-k-1}(S; \mathcal{F}^*)/\beta H_c^{n-k-1}(X; \mathcal{F}^*),$$

where $\mathcal{F}^* = \mathcal{H}om_{\mathcal{O}X}(\mathcal{F}, \Omega_X^n)$, is naturally isomorphic to the topological vector space

$$H^k(X\setminus S; \mathcal{F})/\alpha H^k(X; \mathcal{F}),$$

if and only if the continuous map

$$H_S^{k+1}(X; \mathcal{F}) \longrightarrow H^{k+1}(X; \mathcal{F})$$

has closed kernel.

In particular, if

$$H^k(X; \mathcal{F})=0, \qquad H^{k+1}(X; \mathcal{F})=0,$$

then the topological vector space $\tilde{H}_c^{n-k-1}(S; \mathcal{F}^*)$ is naturally isomorphic to the strong dual to the topological vector space $\tilde{H}^k(X\setminus S; \mathcal{F})$:

$$\{\tilde{H}^k(X\setminus S; \mathcal{F})\}' = \tilde{H}_c^{n-k-1}(S; \mathcal{F}^*).$$

The corollary of the theorem formulated above is nothing but the cohomological analog of the Alexander–Pontryagin topological duality theorem. The latter, as is familiar, establishes a duality between the homology groups, respectively, of a closed set and its complement in Euclidean space (cf. Aleksandrov [5], Pontryagin [1]). In the case of dimension zero, from the Alexander–Pontryagin duality theorem follows the Jordan–Brouwer theorem on the separation of the plane by a closed curve (cf. Aleksandrov [5]). The corollary formulated above reduces, in the corresponding special case, to the Köthe–Silva Dias–Grothendieck duality theorem:

For any open set U on the Riemann sphere S_2, there are natural isomorphisms of topological vector spaces

$$\{\Gamma(U; \mathcal{O})/\alpha\Gamma(S_2; \mathcal{O})\}' = \Gamma(S_2\setminus U; \Omega^1),$$
$$\{\Gamma(U; \Omega^1)\}' = \Gamma(S_2\setminus U; \mathcal{O})/\beta\Gamma(S_2; \mathcal{O})$$

[cf. G. Köthe, *J. Reine Angew. Math.*, **191**, No. 1/2, 29–49 (1953); C. L. da Silva Dias, *Bol. Soc. Mat. São Paulo*, **5**, 1–58 (1950); A. Grothendieck, *J. Reine Angew. Mat.*, **192**, No. 1/2, 35–64, 77–95 (1953)].

An analogous theorem can be proved for functions of several complex variables also:

For any open set U in n-dimensional complex projective space $\mathbf{P}^n$, there are natural isomorphisms of topological vector spaces

$$\{\Gamma(U; \mathcal{O})/\alpha\Gamma(\mathbf{P}^n; \mathcal{O})\}' = \tilde{H}^{n-1}(\mathbf{P}^n\setminus U; \Omega^n),$$

$$\{\Gamma(U; \Omega^1)\}' = H^{n-1}(\mathbf{P}^n\setminus U; \Omega^{n-1})/\beta H^{n-1}(\mathbf{P}^n; \Omega^{n-1})$$

[cf. V. D. Golovin, *Mat. Sb.*, **84**, No. 4, 583–594 (1971)].

8. The map (2) is generally not bijective. Villani [1] proved the following theorem: if the spaces $H^i(X; \mathcal{F})$, $H^i(X_j; \mathcal{F})$ ($i = 0, 1, \ldots, k; j = 1, 2, \ldots$) are separated and $H^i(X_j; \mathcal{F}) = 0$ ($i = k, k + 1, \ldots; j = 1, 2, \ldots$), then $H^k(X; \mathcal{F}) = 0$. Corollary 4.1.2 shows that, in fact, there is a more precise result: if for some k the space $H^k(X; \mathcal{F})$ is separated and $H^k(X_j; \mathcal{F}) = 0$ ($j = 1, 2, \ldots$), then $H^k(X; \mathcal{F}) = 0$ [cf. V. D. Golovin, *Ukr. Geom. Sb.*, **13**, 63–66 (1973)].

9. Theorem 4.2.3 is closely connected with the following classical problem. Let $X_1 \subset X_2 \subset \ldots$ be an increasing sequence of holomorphically complete open sets in the complex space X, which coincides with their union. Is the space X necessarily holomorphically complete? Behnke and Stein [H. Behnke and K. Stein, *Math. Ann.*, **116**, 204–216 (1939)] showed that this is so if X is a domain in the space $\mathbf{C}^n$ (cf. Vladimirov [1, p. 173]). Now, in general the answer is negative: Fornass [1] constructed an example of an increasing sequence (X_j) of open sets in a complex manifold X such that each X_j is isomorphic to a ball in $\mathbf{C}^3$, and $X = \cup X_j$ is not holomorphically complete. Markoe [1] gave a complete solution of the problem. He proved that the following assertions are equivalent:
    a) the space X is holomorphically complete;
    b) (X_j) is a Runge family, i.e., for each compact set K in X there exists a j such that $K \subset X_j$ and the map $\Gamma(X; \mathcal{O}_X) \to \Gamma(X_j; \mathcal{O}_X)$ has everywhere dense image in the topology of uniform convergence on K;
    c) the space $H^1(X; \mathcal{O}_X)$ is separated;
    d) $H^1(X; \mathcal{O}_X) = 0$
    (cf. Markoe [1, pp. 123–124]). Unfortunately, Markoe's proof contains a lacuna; he considers it known that the condition for separatedness of the space $\text{Ext}_{\mathcal{O}_{X,c}}^{-k}(X; \mathcal{F}, K_X)$ given by Ramis, Ruget, and Verdier [1] is not only sufficient but also necessary (cf., however, Theorem 5.1 and Corollary 5.1.3). Since instead of the sheaf $\mathcal{O}_X$ one can obviously consider an arbitrary coherent analytic sheaf $\mathcal{F}$ on X, Markoe's theorem follows directly from Theorem 4.2.3. The more general results of Silva [1] also follow directly as special cases from Theorem 4.2.3.

10. V. P. Palamodov informed the author that Theorem 4.2.3 can also be derived from the Milnor exact sequence [1] (cf. Corollary 4.6 of Chapter 1) with the help of the tests of triviality of the functor $\varprojlim^{(1)}$ [cf. Grothendieck [2, p. 108]; V. P. Palamodov, *Mat. Sb.*, **88**, No. 2, 287–315 (1972)].

11. Directly from Corollary 4.2.4 we get the following: if for each $j = 1$, 2, ... the space $H^k(X_j; \mathscr{F})$ is separated and $H^{k-1}(X_j; \mathscr{F}) = 0$, then the space $H^k(X; \mathscr{F})$ is separated and the map (2) is an isomorphism of topological vector spaces. As a special case we get the following assertion which is usually used in the proof of Cartan's theorem (B), from this (cf., e.g., Andreotti and Vesentini [1]; Gunning and Rossi [1, p. 311]): if $H^k(X_j; \mathscr{F}) = 0$ and $H^{k-1}(X_j; \mathscr{F}) = 0$ for each $j = 1$, 2, ..., then $H^k(X; \mathscr{F}) = 0$.

12. As a corollary of Theorem 4.2.5 we get the following: if for each $j = 1, 2, \ldots$ the image of the space $H^{k-1}X_{j+1}; \mathscr{F})$ is everywhere dense in $H^{k-1}(X_j; \mathscr{F})$ and $H^k(X_j; \mathscr{F}) = 0$, then $H^k(X; \mathscr{F}) = 0$. The results of Silva follow directly from this [cf. A. Silva, *Atti Accad. Naz. Lincei*, **56**, No. 1, 43–44 (1974); *Trans. Am. Math. Soc.*, **199**, 317–326 (1974)].

13. As a corollary of Theorem 4.4.5 we get the following: if for each $j = 1, 2, \ldots$ the space $H_k(X_j; \mathscr{F})$ is separated and $H_{k+1}(X_j; \mathscr{F}) = 0$, then the space $H_k(X; \mathscr{F})$ is separated and the map (5) is an isomorphism of topological vector spaces.

14. We recall that the topological vector space $H^k(X_j; \mathscr{F})$ is separated if and only if the topological vector space $H_{k-1}{}^c(X; \mathscr{F})$ is separated (cf. the corollary to Theorem 2.3 of Chapter 3). Generally the spaces $H^k(X; \mathscr{F})$ are not separated. Serre [2] cited the following example. Let S be a closed set in $\mathbf{C}^2$, which is connected and noncompact, for which $X = \mathbf{C}^2 \setminus S$ is not a domain of holomorphy (for example, S is the real line). Then the topological vector space $H^1(X; \mathscr{O}_X)$ is not separated (cf. Serre [2, pp. 22–23]).

15. See J. Sebastian-i-Silva, *Matematika*, **1**, No. 1, 60–77 (1957); cf. also D. A. Raikov, *Tr. Voronezhsk. Sem. Funkts. Anal.*, **5**, 22–34 (1957).

16. The assertion of Corollary 5.1.3 was first proved in a somewhat different form by Ramis, Ruget, and Verdier [1]. Our proof of Theorem 5.1 is based, essentially, on the same considerations [cf. V. D.

Golovin, *Izv. Akad. Nauk SSSR, Ser. Mat.*, **42**, No. 2, 261–269 (1978)].

17.   The finiteness theorem of Cartan–Serre (cf. point 1.15.2 of Chapter 1) follows easily from Corollary 5.1.3. In fact, let X be a compact complex space. Then by Corollary 5.1.3 the spaces $H^k(X; \mathcal{F})$ are separated. Let $\mathfrak{u} = U_i$ and $\mathfrak{B} = (V_i)$ be finite coverings of the space X by holomorphically complete open sets, where $V_i \subset U_i$ for each i. Then the natural map $C^k(\mathfrak{u}; \mathcal{F}) \to C^k(\mathfrak{B}; \mathcal{F})$ is completely continuous and induces the identity isomorphism of the space $H^k(X; \mathcal{F})$ onto itself. Since this isomorphism is also completely continuous, by Riesz's theorem the space $H^k(X; \mathcal{F})$ is finite-dimensional.

18.   The topological vector space $H_c^k(X; \mathcal{F})$ is separated if and only if the topological vector space $H_{k-1}(X; \mathcal{F})$ is separated (cf. the corollary to Theorem 3.3 of Chapter 3).

19.   The proof of Theorem 5.2 can be based on the following auxiliary assertion [cf. V. D. Golovin, *Usp. Mat. Nauk*, **32**, No. 6, 249–250 (1977)].

Let $E^1 = \lim_{\longrightarrow} E_n^1$, $E^2 = \lim_{\longrightarrow} E_n^2$ be strict inductive limits of strong duals to Frechet–Schwartz spaces and $d: E^1 \to E^2$ be a continuous linear map, for which $d(E_n^1) \subset E_n^2$. Suppose further that $F^1 = \lim_{\longrightarrow} F_n^1$, $F^2 = \lim_{\longrightarrow} F_n^2$ are strict inductive limits of strong duals to Frechet–Schwartz and $\delta: F^1 \to F^2$ is a continuous linear map for which $\delta(F_v^1) \subset F_n^2$. Finally, let $r_1: E^1 \to F^1$, $r_2: E^2 \to F^2$ be continuous linear maps satisfying the following conditions:

a) $r_1(E_n^1) \subset F_n^1$; $r_2(E_n^2) \subset F_n^2$ $(n = 1, 2, \ldots)$;

b) $\delta \circ r_1 = r_2 \circ d$;

c) the maps $r_1: E_n^1 \to F_n^1$ $(n = 1, 2, \ldots)$ are completely continuous;

d) the map $E^2/d(E^1) \to F^2/\delta(F^1)$ induced by the map r_2 is injective.

Then the image $d(E^1)$ is closed in the space E^2 if and only if the following condition holds: for each positive integer n and each bounded set $B \subset E_n^2$ there exists an integer $m \geq n$ such that

$$B \cap d(E^1) \subset d(E_m^1).$$

In weaker form the assertion cited was used previously by other authors for analogous goals (cf. Malgrange [4], Ramis, Ruget, and Verdier [1], Andreotti and Banica [1]).

20. The assertion of Corollary 5.2.3 was first proved by Ramis, Ruget, and Verdier [1]. A simpler proof, based on the same considerations, was given by Andreotti and Banica [1].

Let X be a holomorphically convex complex space and $\mathscr{F}$ be a coherent analytic sheaf on X. Ramis, Ruget, and Verdier [1] proved, with the help of Corollary 5.2.3, that the topological vector spaces $H_c^k(X; \mathscr{F})$ [and hence also the spaces $H_k(X; \mathscr{F})$] are separated. Analogously, with the help of Corollary 5.1.3 one can prove that under the assumptions made the topological vector spaces $H^c{}_k(X; \mathscr{F})$ and $H^k(X; \mathscr{F})$ are also separated (cf. Ramis [3]). On the other hand, Malgrange [8] showed that there exists a pseudoconvex manifold X, obtained by removing a real hyperplane from a two-dimensional complex torus ("Grauert's example"; cf., e.g., Andreotti [2, pp. 97–99]), such that the topological vector space $H^1(X, \mathcal{O}_X)$ is not separated.

Chapter 5

1. In connection with the definition of the dimension of an arbitrary set in a complex space, cf. also Scheja [1, p. 350].

2. Let $\mathscr{F}$ be a coherent analytic sheaf on the complex space X and S be an analytic set of dimension d in X. Reiffen [1] proved that then

$$H_c^k(S; \mathscr{F}) = 0 \quad \text{for} \quad k > d.$$

In the special case considered, Theorem 1.1 can be derived from this with the help of the duality theorem of 3.8 of Chapter 4.

3. Let $\mathscr{F}$ be a coherent analytic sheaf on the complex space X and $\mathscr{J}$ be the largest subsheaf of ideals in $\mathcal{O}_X$, which has the property that $\mathscr{J}\mathscr{F} = 0$. Then the support Supp $\mathscr{F}$ coincides with the set of those $x \in X$, for which $\mathscr{J}_x \neq \mathcal{O}_{X,x}$. Since the sheaf $\mathscr{J}$ is coherent, Supp $\mathscr{F}$ is an analytic set in X (cf. Fischer [1, p. 8]).

4. Corollary 1.1.3 is the homological analog of a theorem of Reiffen [1, p. 275] (cf. Note 2).

Let X be a complex manifold of dimension n. Then for any analytic sheaf $\mathscr{F}$ on X, from Corollary 1.1.3 we get

$$\operatorname{Ext}^{n-k}_{\mathcal{O}_X, \, \Phi}(X; \, \mathscr{F}, \, \Omega^n_X) = 0 \quad \text{for} \quad k > d,$$

where d is the dimension of the support Supp $\mathscr{F}$ (cf. Golovin [2, p. 753]). In particular, a result of Kerner [1] follows:

$$\mathscr{E}\mathrm{xt}^{n-k}_{\mathcal{O}_X}(\mathscr{F}, \, \mathcal{O}_X) = 0 \quad \text{for} \quad k > d.$$

5.   The assertion of Corollary 1.2.1 is a familiar theorem of Malgrange [1] and Siu [2]. The proof recounted in the text is due essentially to Ramis, Ruget, and Verdier [1, pp. 281–282].

6.   The concept of homological codimension of a coherent analytic sheaf was first introduced by Andreotti and Grauert [1] (cf. p. 110 of the Russian translation). Essentially it coincides with the concept of depth of a coherent analytic sheaf prof $\mathscr{F}$ which arose originally in local algebra (cf. Serre [5] and Banica and Stanasila [4]).

7.   The historically first and most classical application of duality theorems for cohomology spaces of coherent analytic sheaves is the proof due to Serre of the Riemann–Roch theorem on Riemann surfaces (cf. Serre [2, pp. 24–25]; Andreotti [2, pp. 143–150]; Grauert and Remmert [2, pp. 225–232]). As applications of duality theorems for cohomology spaces of coherent analytic sheaves, Serre also found the following results. First, a generalization of a classical theorem of Osgood (cf. Vladimirov [1, p. 180]): if X is a holomorphically complete complex manifold of dimension $n \geq 2$, K is an arbitrary compact subset of X, and f is a holomorphic function on $X \setminus K$, then there exists a holomorphic function g on X, which coincides with f outside a compact set $K' \supset K$ (cf. Serre [1, p. 370]). Second, the solvability of the first Cousin problem (cf. Gunning and Rossi [1, p. 317]): if X is a holomorphically complete complex manifold of dimension $n \neq 2$, and U is a relatively compact holomorphically complete open set in X, then over the set $S = X \setminus U$ the first Cousin problem is solvable (cf. Serre [1, p. 370]). Banica and Stanasila [3] generalized these results of Serre with the help of the concept of homological codimension.

8.   Banica and Stanasila [1–3] found the following more complete result:
Let X be a holomorphically complete complex space, $\mathscr{F}$ be an integer. Then the following assertions are equivalent:
a) codh $\mathscr{F} \geq h$;

b) for any compact set S in X, for which there exists a fundamental system of holomorphically complete open neighborhoods, $H_s^k(X; \mathscr{F}) = 0$ for $k < h$;

c) $H_c^k(X; \mathscr{F}) = 0$ for $k < h$.

9. In the quite special case when X is a domain of the space $\mathbf{C}^n$, and $\mathscr{F} = \mathscr{O}_X$ is its structure sheaf, the assertion of the corollary to Theorem 2.1 was proved by Friedman [A. Friedman, *Bull. Am. Math. Soc.*, **72**, No. 3, 505–507 (1966)].

10. Theorem 2.2 was first proved by the author for complex manifolds [cf. V. D. Golovin, *Funkts. Anal. Ego Prilozhen.*, **5**, No. 5, 66 (1971); *Teor. Funkts., Funkts. Anal. Ikh Prilozhen.*, **16**, 74–78 (1972)]. The assertion of the corollary to Theorem 2.2 was first proved by Scheja [1] with the help of Riemann's theorem on the continuation of holomorphic functions (cf., e.g., Hervé [1, pp. 55–56]). Obviously the latter is a special case (for $k = 0$ and $\mathscr{F} = \mathscr{O}_X$) of the corollary to Theorem 2.2.

11. The concept of strongly p-convex function was first introduced by Rothstein [W. Rothstein, *Math. Ann.*, **129**, No. 1, 96–138 (1955)]. Strongly 1-convex functions are smooth strictly plurisubharmonic functions (cf., e.g., Hörmander [1, p. 70]). The concept of strongly p-convex function was generalized essentially by de Graeve [1] (cf. Malgrange [4]).

The class of strongly p-convex (respectively q-concave) complex spaces was first introduced and studied by Andreotti and Grauert (cf. Grauert [3] and Andreotti and Grauert [1]). This class is rather large: it includes both compact and holomorphically complete spaces; more precisely, strong 0-convex spaces are compact spaces, and strongly 1-convex spaces are complex spaces of finite type, i.e., are obtained from holomorphically complete spaces by blowing up a finite number of points (cf. Narasimhan [1]).

The work of Andreotti and Grauert is essentially the development of the ideas and methods which arise in connection with the solution of Levi's problem for strongly 1-convex complex spaces (Grauert [2] and Narasimhan [1]). Analogous results for locally free sheaves on complex manifolds were found by Andreotti and Vesentini [2] with the help of the methods of the theory of differential forms. A new approach to the theory of strongly p-convex (respectively strongly q-concave) complex spaces, based on ideas of Malgrange [4], was suggested by de Graeve [1] and Ramis [1-3]. Using the duality theorems and separation tests for cohomology spaces of coherent analytic sheaves systematically, Ramis [3] found the results of Andreotti and

Grauert in stronger and definitive form (cf. the survey of Onishchik [2, pp. 119–120]).

12.   Theorem 3.1 and the corollary to Theorem 3.4 are essentially the basic results of Andreotti and Grauert [1].

13.   Let X be a strongly p-convex complex space of dimension n, $\mathscr{L}$ be a locally free analytic sheaf of finite range Rg $\mathscr{L}$, and $\mathscr{F}$ be a coherent analytic sheaf on X. Then, as Leistner [1] showed, for $k \geq p$ one has the estimate

$$\dim_C H^k (X; \mathscr{F} \otimes \mathscr{L}) \leqslant A \,(\mathrm{Rg}\,\mathscr{L}) \,(\log \| g \| + B)^n,$$

where $g = (g_{ij}) \in Z^1(\mathfrak{u}; \mathcal{O}_X{}^*)$ is a cocycle defining the sheaf $\mathscr{L}$ relative to a finite family $\mathfrak{u} = (U_i)$ of open sets in X, $\|g\| = \max \|g_{ij}\|$, and A and B are constants, independent of $\mathscr{L}$. This same estimate holds for a strongly q-concave complex space X for $k < \mathrm{codh}\,\mathscr{F} - q$.

14.   The assertion of the corollary to Theorem 3.2 was first proved by Andreotti and Kas [2] for complex manifolds.

15.   Let X be a complex space, $\mathscr{L}$ be a fixed invertible analytic sheaf, and $\mathscr{F}$ be an arbitrary coherent analytic sheaf on X. Andreotti and Banica [2] considered the twisted sheaves $\mathscr{F}(m) = \mathscr{F} \otimes \mathscr{L}^m$ (m being an integer) and proved vanishing theorems for their cohomology groups. For example, if X is a strongly p-convex space of finite dimension and $\mathscr{L} > 0$, then for sufficiently large m,

$$H^k (X; \mathscr{F}(m)) = 0 \quad \text{for} \quad k \geqslant p,$$
$$H^k_c (X; \mathscr{F}(-m)) = 0 \quad \text{for} \quad k \leqslant \mathrm{codh}\,\mathscr{F} - p.$$

If X is a strongly q-concave space of finite dimension and $\mathscr{L} < 0$, then for sufficiently large m,

$$H^k (X; \mathscr{F}(m)) = 0 \quad \text{for} \quad k < \mathrm{codh}\,\mathscr{F} - q,$$
$$H^k_c (X; \mathscr{F}(-m)) = 0 \quad \text{for} \quad k > q,$$

where it is assumed that $\mathscr{F}$ is a Cohen–Macaulay sheaf.

16.   The sheaves of germs of holomorphic differential forms Ω_X^p on complex spaces were first defined by Grauert [H. Grauert, *Math. Ann.*, **146**, No. 4, 331–368 (1962); cf. also Grauert and Kerner [1] and Reiffen [2]].

17. The complex $\Omega_X^\bullet$ on complex spaces is not generally a resolution of the simple sheaf C; in other words, the Poincaré lemma may not hold on complex spaces (cf. Reiffen [2] and Bloom and Herrera [1]).

18. The hypercohomology groups $H_\Phi{}^k(X; \Omega_X^\bullet)$ are usually denoted by $H_{DR,\Phi}{}^k(X; C)$ (cf. Bloom and Herrera [1, p. 288]). If X is a complex manifold, then there is a natural isomorphism

$$H^k_{DR,\Phi}(X; C) = H^k_\Phi(X; C)$$

(the "holomorphic de Rham theorem"; cf. Serre [1, p. 363]). In general, when X is an arbitrary complex space, the cited isomorphism is replaced by the Bloom–Herrera decomposition.

19. Theorem 4.3 was proved by the author [cf. V. D. Golovin, *Usp. Mat. Nauk*, **37**, No. 5, 177–178 (1982); *Izv. Vyssh. Uchebn. Zaved., Mat.*, No. 2, 11–15 (1982)].

20. Let M be an algebraic variety of dimension m, imbedded without singularities in a complex projective space, and let S be the section of the variety M by a hyperplane in general position. Then according to the Lefschetz theorem on hyperplane sections, the restriction map

$$H^k(M; C) \longrightarrow H^k(S; C) \qquad (*)$$

is bijective for $k < m - 1$ and injective for $k = m - 1$ (cf. Lefschetz [1, p. 89] and Chern [1, p. 146]). The proof of this assertion can be based on the following considerations. Let X be a holomorphically complete complex manifold of dimension n. Then by a theorem of Serre

$$H^c_k(X; C) = 0 \quad \text{for} \quad k > n$$

(cf. Serre [1, p. 364]). Consequently, by Poincaré duality

$$H^k_c(X; C) = 0 \quad \text{for} \quad k < n.$$

The Lefschetz theorem is obtained from this with the help of the cohomology exact sequence connected with the closed subset S if one sets $X = M \setminus S$ (cf. Narasimhan [2]). Many authors concerned themselves with generalizing the Lefschetz theorem in various directions [cf., e.g., Kodaira and Spencer [1], Andreotti and Frankel [1], L. Kaup, *Sitzungsber. Bayer Akad. Wiss., Math.-Naturwiss. Kl., 1966*, München (1967), pp. 163–165), W. Barth, *Am. J. Math.*, **92**, No. 4, 951–967 (1970), M. Schneider, *Invent. Math.*, **31**, No. 2, 183–192

(1975), H. Jamaguchi, *Comment. Math. Univ. St. Pauli*, **26**, No. 2, 157–160 (1978)]. However, since Poincaré duality does not hold in general on analytic sets with singularities, a satisfactory generalization of the Lefschetz theorem to those cases when the analytic set $M \setminus S$ contains singularities was not found. On the other hand, as a special case of Corollary 4.4 we get the following generalization of the Lefschetz theorem on hyperplane sections, which is suitable for analytic sets with any singularities:

Let M be an analytic set in complex projective space, S be a section of the set M by a hyperplane which does not contain M, and $\rho = \rho(M \setminus S)$. Then the restriction map (*) is bijective for $k < p - 1$, and injective for $k = \rho - 1$.

In particular, if the analytic set $M \setminus S$ has no singularities, then ρ is equal to the dimension of $M \setminus S$, so that in this case we get the Lefschetz theorem in its original form (cf. Chern [1, p. 146]).

21. The concept of a real analytic space and of its complexification were introduced by Cartan [6].

22. The space $\mathbf{C}^n$ is a complexification of the space $\mathbf{R}^n$ under the natural inclusion (cf. Cartan [6]). If X is a complex space considered as real analytic, then a complexification for X is the product $X \times \overline{X}$, while X is then isomorphic as a ringed space to the "diagonal" of this product [cf. V. D. Golovin, *Mat. Zametki*, **9**, No. 5, 569–573 (1971)].

23. Let X be a real analytic space which is countable at infinity, and $\widetilde{X}$ be a complexification of it. One can assume that the space $\widetilde{X}$ is countable at infinity, X is a closed set in $\widetilde{X}$, and the sheaf $\mathscr{A}_X$ is induced on X by the sheaf $\mathcal{O}_{\widetilde{X}}$. Grauert [2] showed that for any open set $U \subset X$ in the space $\widetilde{X}$ there exists a fundamental system of holomorphically complete open neighborhoods $\widetilde{U}$ such that $U = \widetilde{U} \cap X$. By Cartan's theorem (B) then

$$H^k(U; \mathscr{F}) = 0 \quad \text{for} \quad k \neq 0$$

for any coherent analytic sheaf $\mathscr{F}$ on $\widetilde{X}$ (cf. Cartan [6]).

24. It follows from Lemma 5.2 that the contravariant functor $\mathscr{F} \mapsto H_0^X(\widetilde{U}; \mathscr{F})$, defined on the category of coherent analytic sheaves on

$\widetilde{X}$, and assuming values in the category of complex vector spaces, is exact.

25. It follows in particular from Theorem 5.4 that the presheaf $\widetilde{U} \mapsto H_0{}^X(\widetilde{U}; \mathscr{F})$ is defined by the sheaf $\mathscr{H}_0{}^X(\mathscr{F})$ (cf. point 1.2.1 of Chapter 1).

26. Definition 5.6 was first introduced by the author [cf. Golovin [7], and also V. D. Golovin, *Vestn. Khar'kovsk. Univ.*, No. 230, 82–88 (1982)]. The new approach to the theory of hyperfunctions based on this definition (and on the homology theory of analytic sheaves) has a number of obvious advantages, since it permits one to get all the basic properties of hyperfunctions, including the theorem that the sheaf of germs of hyperfunctions is an injective module over the sheaf of germs of analytic functions (Theorem 5.9) in a simpler and more general situation (i.e., on real analytic spaces).

27. Let X be a real analytic space which is countable at infinity and $\widetilde{X}$ be its complexification. Then for any closed set $S \subset X$ there is a natural isomorphism

$$\Gamma_S(X; \mathscr{B}_X) = H_0^S(\widetilde{X}; \mathcal{O}_{\widetilde{X}})$$

(as can be proved analogously to Theorem 5.4). If S is a compact set in X, then by Theorem 3.9 of Chapter 4 one gets a natural isomorphism of vector spaces

$$\Gamma_S(X; \mathscr{B}_X) = \{\Gamma(S; \mathcal{O}_{\widetilde{X}})\}',$$

i.e., the vector space $\Gamma_S(X; \mathscr{B}_X)$ can be identified with the space of analytic functionals on S, since $\mathcal{O}_{\widetilde{X}}/X = \mathscr{A}_X$ is the sheaf of germs of (real) analytic functions on X (by the definition of complexification). Thus, if $X = R^n$, Definition 5.6 agrees with the definition of hyperfunctions by Martineau [1] (cf. Schapira [1, p. 62]).

If X is a real analytic manifold of dimension n, then $\widetilde{X}$ is a complex manifold of (complex) dimension n. By Poincaré duality (cf. Corollary 4.2 of Chapter 2) for an arbitrary open set $U \subset X$ there is a natural isomorphism

$$\Gamma(U; \mathscr{B}_X) = H_U^n(\widetilde{U}; \Omega_{\widetilde{X}}^n),$$

which is equivalent with the definition of hyperfunctions by Sato [1] (cf. Schapira [1, p. 115]).

28. Theorem 5.8 is usually called *the hyperfunction division theorem,* and was first proved for real analytic manifolds by Kantor [1] (cf. Schapira [1, p. 107]).

29. Theorem 5.9 was original proved by the author for hyperfunctions in domains of complex number space [cf. V. D. Golovin, *Mat. Zemetki,* **18**, No. 4, 589–596 (1975)], and later in general (cf. Golovin [7]).

30. From Theorem 5.9 we get the following improvement of the division theorem for hyperfunctions (i.e., Theorem 5.8) directly:

Suppose given for each open set $U \subset X$ a hyperfunction $\alpha_U \in \Gamma(U; \mathscr{B}_X)$ and an analytic function $f_U \in \Gamma(U; \mathscr{A}_X)$. Let us assume that each relation of the form

$$\theta_1 f_{U_1} + \ldots + \theta_p f_{U_p} = 0$$

with analytic $\theta_1, \ldots, \theta_p$ implies the relation

$$\theta_1 \alpha_{U_1} + \ldots + \theta_p \alpha_{U_p} = 0.$$

Then there exists a hyperfunction $\alpha \in \Gamma(X; \mathscr{B}_X)$ such that $f_U \alpha = \alpha_U$ for each U (cf. Note 11 to Chapter 1).

31. Theorem 5.11 was first proved for real analytic manifolds by Schapira [2].

REFERENCES

P. S. Aleksandrov

1. "Determination of the Betti numbers of an arbitrary closed set," in: P. S. Aleksandrov, *General Homology Theory* [in Russian], Nauka, Moscow (1979), pp. 56–69.
2. "Investigation of the form and disposition of closed sets of arbitrary dimension," *ibid.*, pp. 68–175.
3. "Betti groups and homology ring of a locally compact space," *Dokl. Akad. Nauk SSSR*, **26**, No. 7, 632–634 (1940).
4. "General homology theory," *Uch. Zap. Mosk. Gos. Univ. im. M. V. Lomonosova*, **45**, 3–60 (1940).
5. *Combinatorial Topology* [in Russian], Gostekhizdat, Moscow–Leningrad (1947).

A. Andreotti

1. "Theorèms de dépendence algébrique sur les espaces complexes pseudoconcaves," *Bull. Soc. Math. Fr.*, **91**, No. 1, 1–38 (1963).
2. *Nine Lectures on Complex Analysis*, Ed. Cremonese, Roma (1974).

A. Andreotti and C. Banica

1. "Relative duality on complex spaces," *Rev. Roum. Math. Pures Appl.*, **20**, No. 9, 981–1041 (1975); **21**, No. 9, 1139-1181 (1976).
2. "Twisted sheaves on complex spaces," *Ann. Scu. Norm. Sup. Pisa*, **7**, No. 1, 1–27 (1980).

A. Andreotti and T. Frankel

1. "The Lefschetz theorem on hyperplane sections," *Ann. Math.*, **69**, No. 3, 713–717 (1959).

A. Andreotti and H. Grauert

1. "Théorèmes de finitude pour la cohomologie des espaces complexes," *Bull. Soc. Math. Fr.*, **90**, No. 2, 193–259 (1962).

A. Andreotti and A. Kas

1. "Serre duality on complex analytic spaces," *Atti Accad. Naz. Lincei*, **50**, No. 4, 397–401 (1971).
2. "Duality on complex spaces," *Ann. Scu. Norm. Sup. Pisa*, **27**, No. 2, 187–263 (1973).

A. Andreotti and E. Vesentini

1. "Les théorèmes fondamentaux de la théorie des espaces holomor-phiquement complets," *Sem. Ch. Ehresmann, Paris*, **4**, 1–31 (1962–1963).
2. "Carleman estimates for the Laplace–Beltrami equations on complex manifolds," *Inst. Hautes Étud. Sci., Publ. Math.*, No. 25, 313–362 (1965).

C. Banica

1. "Un théorème concernant la séparation de certains espaces de coho-mologie," *C. R. Acad. Sci. Paris*, **272**, No. 12, 782–785 (1971).

C. Banica and O. Stanasila

1. "Sur la profondeur d'un faisceau analytique cohérent sur un espace de Stein," *C. R. Acad. Sci. Paris*, **269**, No. 15, 636–639 (1969).
2. "Sur la cohomologie des faisceaux analytiques cohérents à support dans un compact homomorphe-convexe," *C. R. Acad. Sci. Paris*, **270**, No. 18, 1174–1177 (1970).
3. "Some results on the extension of analytic entities defined out of a compact," *Ann. Scu. Norm. Sup. Pisa*, **25**, No. 2, 347–376 (1971).
4. *Metode Algebrice în Teoria Globala a Spatiilor Complexe*, Ed. Acad. RSR, Bucuresti (1974).

T. Bloom and M. Herrera

1. "De Rham cohomology of an analytic space," *Invent. Math.*, **7**, No. 4, 275–296 (1969).

N. Bourbaki

1. *General Topology. Fundamental Structures* [Russian translation], Fizmatgiz, Moscow (1958).
2. *Topological Vector Spaces* [Russian translation], IL, Moscow (1959).

H. Cartan

1. "Idéaux et modules de fonctions analytiques de variables complexes," *Bull. Soc. Math. Fr.*, **78**, No. 1, 28–64 (1950).
2. *Cohomologie des Groupes, Suite Spectrale, Faisceaux*, Séminaire E. N. S., Paris, 1950–1951.
3. *Théorie des Fonctions de Plusieurs Variables*, Séminaire E. N. S., Paris, 1951–1952.
4. "Variétés analytiques complexes et cohomologie," in: *Coll. Fonct. Plus. Variables, Bruxelles*, Masson, Paris (1953), pp. 41–55.
5. *Théorie des Fonctions Automorphes et des Espaces Analytiques*, Séminaire E. N. S., Paris, 1953–1954.
6. "Variétés analytiques réeles et variétés analytiques complexes," *Bull. Soc. Math. Fr.*, **85**, No. 1, 77–99 (1957).
7. "Sur les fonctions de plusieurs variables complexes: les espaces analytiques," in: *Proc. International Congress of Mathematicians, 1958*, Cambridge University Press, Cambridge (1960), pp. 33–52.

H. Cartan and S. Eilenberg

1. *Homological Algebra* [Russian translation], IL, Moscow (1960).

H. Cartan and J.-P. Serre

1. "Un théorème de finitude concernant les variétés analytiques compactes," *C. R. Acad. Sci. Paris*, **237**, No. 2, 128–130 (1953).

S. S. Chern

1. *Complex Manifolds* [Russian translation], IL, Moscow (1961).

R. Deheuvels

1. "Homologie à coefficients dans un antifaisceau," *C. R. Acad. Sci. Paris*, **250**, No. 14, 2492–2494 (1960).

2. "Homologie des ensembles ordonnés et des espaces topologiques," *Bull. Soc. Math. Fr.*, **90**, No. 2, 261–321 (1962).

P. Dolbeault

1. "Sur la cohomologie des variétés analytiques complexes," *C. R. Acad. Sci. Paris*, **236**, No. 2, 175–177 (1953).

B. Eckmann and A. Schopf

1. "Über injektive Moduln," *Arch. Math.*, **4**, No. 2, 75–78 (1953).

G. Fischer

1. *Complex Analytic Geometry*, Springer-Verlag, Berlin–New York (1976).

J. E. Fornass

1. "An increasing sequence of Stein manifolds whose limit is not Stein," *Math. Ann.*, **223**, No. 3, 275–277 (1976).

O. Forster and K. Knorr

1. "Ein Beweis des Grauertschen Bildgarbensatzes nach Ideen von B. Malgrange," *Manuscr. Math.*, **5**, No. 1, 19–44 (1971).

F. Fouché

1. "Un complexe dualisant en géométrie analytique complexe," *C. R. Acad. Sci. Paris*, **280**, No. 17, 1141–1143 (1975).
2. "Un complexe dualisant en géométrie analytique," Séminaire F. Norguet, *Lect. Notes Math.*, **482**, 282–332 (1975).

R. Godement

1. *Algebraic Topology and Sheaf Theory* [Russian translation], IL, Moscow (1961).

V. D. Golovin

1. "Duality for coherent analytic sheaves," *Dokl. Akad. Nauk SSSR*, **191**, No. 4, 755–758 (1970).
2. "Duality for cohomology with compact supports," *Dokl. Akad. Nauk SSSR*, **199**, No. 4, 751–753 (1971).

3. "Global dimension of the sheaf of germs of holomorphic functions," *Dokl. Akad. Nauk SSSR*, **223**, No. 2, 273–275 (1975).
4. "Homology of analytic sheaves," *Dokl. Akad. Nauk SSSR*, **225**, No. 1, 41–43 (1975).
5. "Homological properties of fine sheaves," *Soobshch. Akad. Nauk GSSR*, **83**, No. 3, 565–567 (1976).
6. "Duality theorems on complex spaces," *Soobshch. Akad. Nauk GSSR*, **91**, No. 1, 21–24 (1978).
7. "Hyperfunctions on analytic spaces," *Soobshch. Akad. Nauk GSSR*, **95**, No. 1, 21–23 (1979).
8. "Relative version of the Oka–Cartan–Serre theory," *Dokl. Akad. Nauk SSSR*, **260**, No. 1, 17–19 (1981).

R. de Graeve

1. "Ouverts strictement *p*-convexes d'un espace analytique complexe," *C. R. Acad. Sci. Paris*, **277**, No. 5, 243–245 (1973).

H. Grauert

1. "Charakterisierung der holomorphvollständigen komplexen Räume," *Math. Ann.*, **129**, No. 3, 233–259 (1955).
2. "On Levi's problem and the imbedding of real-analytic manifolds," *Ann. Math.*, **68**, No. 2, 460–472 (1958).
3. "Une notion de dimension cohomologique dans la théorie des espaces complexes," *Bull. Soc. Math. Fr.*, **87**, No. 4, 341–350 (1959).
4. "Ein Theorem der analytischen Garbentheorie und die Modulräume komplexer Strukturen," *Inst. Hautes Études Sci., Publ. Math.*, No. 5, 1–64 (1960); No. 16, 131–132 (1963).

H. Grauert and H. Kerner

1. "Deformationen von Singularitäten komplexer Räume," *Math. Ann.*, **153**, No. 3, 236–260 (1964).

H. Grauert and R. Remmert

1. "Bilder und Urbilder analytischer Garben," *Ann. Math.*, **68**, No. 2, 393–443 (1958).
2. *Theory of Stein Spaces*, Springer-Verlag, Berlin–New York (1979).

A. Grothendieck

1. "Théorèmes de dualité pour les faisceaux algébrique cohérents," *Sém. Bourbaki*, **9**, No. 149 (1956–1957).

2. *Questions of Homological Algebra* [Russian translation], IL, Moscow (1961).
3. "The cohomology theory of abstract algebraic varieties," in: *Proc. International Congress of Mathematicians, 1958*, Cambridge Univ. Press, Cambridge (1960), pp. 103–118.
4. *Local Cohomology*, Springer-Verlag, Berlin–New York (1967).

R. C. Gunning and H. Rossi

1. *Analytic Functions of Several Complex Variables* [Russian translation], Mir, Moscow (1969).

R. Hartshorne

1. *Residues and Duality*, Springer-Verlag, Berlin–New York (1966).
2. *Algebraic Geometry* [Russian translation], Mir, Moscow (1981).

M. Hervé

1. *Functions of Several Complex Variables* [Russian translation], Mir, Moscow (1965).

H. Hironaka

1. "Resolution of singularities of an algebraic variety over a field of characteristic zero," *Ann. Math.*, **79**, No. 1, 109–180 (1964).

L. Hörmander

1. *Introduction to the Theory of Functions of Several Complex Variables* [Russian translation], Mir, Moscow (1968).

J.-M. Kantor

1. "Hyperfonctions cohérentes," *C. R. Acad. Sci. Paris*, **269**, No. 1, 18–20 (1969).

H. Kerner

1. "Kohärente analytische Garben mit niederdimensionalen Träger," *Sitzungsber. Bayer. Akad. Wiss., 1966*, pp. 41–51 (1967).

R. Kiehl and J.-L. Verdier

1. "Ein einfacher Beweis des Kohärenzsatzes von Grauert," *Math. Ann.*, **195**, No. 1, 24–50 (1971).

K. Knorr

1. "Der Grauertsche Projektionssatz," *Invent. Math.*, **12**, No. 2, 118–172 (1971).

K. Kodaira and D. C. Spencer

1. "On a theorem of Lefschetz and the lemma of Enriques–Severi–Zariski," *Proc. Natl. Acad. Sci. USA*, **39**, No. 12, 1273–1278 (1953).

H. Laufer

1. "On Serre duality and envelopes of holomorphy," *Trans. Am. Math. Soc.*, **128**, No. 3, 414–436 (1967).

S. Lefschetz

1. *"L'Analysis Situs et la Géométrie Algébrique*, Gauthiers-Villars, Paris (1924).
2. *Algebraic Topology*, American Mathematical Society, Providence, Rhode Island (1942).

D. Leistner

1. "Cohomologietheorie mit Schränken für pseudokonvexe und pseudokonkave komplexe Räume," *J. Rein. Angew. Math.*, No. 311/312, 315–329 (1979).

J. Leray

1. "L'anneau spectral et l'anneau filtré d'homologie d'un espace localement compact et d'une application continue," *J. Math. Pures Appl.*, **29**, No. 1–2, 1–139 (1950).

S. MacLane

1. *Homology* [Russian translation], Mir, Moscow (1966).

B. Malgrange

1. "Faisceaux sur les variétés analytiques-réels," *Bull. Soc. Math. Fr.*, **85**, No. 2, 231–237 (1957).
2. *Lectures on the Theory of Functions of Several Complex Variables* [Russian translation], Nauka, Moscow (1969).
3. "Systèmes différentiels à coefficients constants," *Sém. Bourbaki*, **15**, No. 246 (1962–1963).
4. "Some remarks on the convexity for differential operators," in: *Differential Analysis. Bombay Colloquium*, Oxford University Press, Oxford (1964), pp. 163–174.
5. *Ideals of Differentiable Functions* [Russian translation], Mir, Moscow (1968).
6. "Analytic spaces," *Enseign. Math.*, **14**, No. 1, 1–28 (1968).
7. "Ouverts concaves et théorèmes de dualité pour les systèmes différentiels à coefficients constants," *Commun. Math. Helv.*, **46**, No. 4, 487–499 (1971).
8. "La cohomologie d'une variété analytique complexe à bord pseudoconvexe n'est pas nécessairement séparée," *C. R. Acad. Sci. Paris*, **280**, No. 3, 93–95 (1975).

A. Markoe

1. "Runge families and inductive limits of Stein spaces," *Ann. Inst. Fourier*, **27**, No. 3, 117–127 (1977).
2. "A holomorphically convex analog of Cartan's theorem B," *Ann. Math. Stud.*, No. 100, 291–298 (1981).

A. Martineau

1. "Les hyperfonctions de M. Sato," *Sém. Bourbaki*, **13**, No. 214 (1960-1961).

J. Milnor

1. "On axiomatic homology theory," *Pac. J. Math.*, **12**, No. 1, 337–341 (1962).

R. Narasimhan

1. "The Levi problem for complex spaces," *Math. Ann.*, **142**, No. 4, 355–365 (1961); **146**, No. 3, 195–216 (1962).
2. "Compact analytical varieties," *Enseign. Math.*, **14**, No. 1, 75–98 (1968).

K. Oka

1. "Sur les fonctions analytiques de plusieurs variables, VII: Sur quelques notions arithmetiques," *Bull. Soc. Math. Fr.*, **78**, No. 1, 1–27 (1950).

A. L. Onishchik

1. "Stein spaces," *Itogi Nauki Tekh., Ser. Alg., Topol., Geom.*, **11**, 125–151 (1974).
2. "Pseudoconvexity in the theory of complex spaces," *Itogi Nauki Tekh., Ser. Alg., Topol., Geom.*, **15**, 93–171 (1977).

L. S. Pontryagin

1. "The general topological theorem of duality for closed sets," *Ann. Math.*, **35**, No. 4, 904–914 (1934).

J.-P. Ramis

1. "Théorèmes de dualité en géométrie analytique complexe et applications," *Colloq. Int. C.N.R.S.*, No. 208, 211–220 (1974).
2. "Théorèmes de finitude pour les espaces p-convexes, q-concaves et (p, q)-convexes-concaves," Séminaire F. Norquet, *Lect. Notes Math.*, **409**, 357–393 (1974).
3. "Théorèmes de separation et de finitude pour l'homologie et la cohomologie des espaces (p, q)-convexes-concaves," *Ann. Scu. Norm. Sup. Pisa*, **27**, No. 4, 933–997 (1974).
4. "Géométrie analytique et géométrie algébrique (variations sur le théme 'GAGA'," Séminaire P. Lelong-H. Skoda, *Lect. Notes Math.*, **694**, 228–289 (1978).

J.-P. Ramis and G. Ruget

1. "Complexe dualisant et théorèmes de dualité en géométrie analytique complexe," *Inst. Hautes Étud. Sci., Publ. Math.*, No. 38, 77–91 (1970).
2. "Dualité relative et images directes en géométrie analytique," *C. R. Acad. Sci. Paris*, **276**, No. 12, 843–845 (1973).
3. "Résidues et dualité," *Invent. Math.*, **26**, No. 2, 89–131 (1974).

J.-P. Ramis, G. Ruget, and J.-L. Verdier

1. "Dualité relative en géométrie analytique complexe," *Invent. Math.*, **13**, No. 4, 261–283 (1971).

H.-J. Reiffen

1. "Riemannsche Hebbarketissätze für Cohomologieklassen mit kompaktem Träger," *Math. Ann.*, **164**, No. 3, 272–279 (1966).
2. "Das Lemma von Poincare für holomorphe Differentialformen auf komplexen Räumen," *Math. Z.*, **101**, No. 4, 269–284 (1967).

G. de Rham

1. *Differentiable Manifolds* [Russian translation], IL, Moscow (1956).

J.-E. Roos

1. "Sur les foncteurs dérivés de $\varprojlim$ Applications," *C. R. Acad. Sci. Paris*, **252**, No. 24, 3702–3704 (1961).

G. Ruget

1. "Complexe dualisant et résidus," *Bull. Soc. Math. Fr., Suppl.*, No. 38, 31–34 (1974).

M. Sato

1. "Theory of hyperfunctions," *J. Fac. Sci. Univ. Tokyo, Sec. I*, **8**, No. 1, 139–193 (1959); **8**, No. 2, 387–437 (1960).

P. Schapira

1. *Theory of Hyperfunctions* [Russian translation], Mir, Moscow (1972).
2. "Utilisation des hyperfonctions dans les théorèmes de dualité de la géométrie analytique," Séminaire P. Lelong, *Lect. Notes Math.*, **205**, 166–182 (1971).

G. Scheja

1. "Riemannsche Hebbarkeitssätze für Cohomologieklasses," *Math. Ann.*, **144**, No. 4, 345–360 (1961).
2. *Complex Analytic Manifolds. Elliptic Partial Differential Equations* [Russian translation], Mir, Moscow (1964).

L. S. Schwartz

1. "Homomorphismes et applications complétement continues," *C. R. Acad. Sci. Paris*, **236**, No. 26, 2472–2473 (1953)

2. *Complete Analytic Manifolds. Elliptic Partial Differential Equations* [Russian translation], Mir, Moscow (1964).

J.-P. Serre

1. "Quelques problèmes globaux relatifs aux variétés de Stein," in: *Coll. Fonct. Plus Variabl. Bruxelles*, Masson, Paris (1953), pp. 57–68.
2. "Un théorème de dualité," *Commun. Math. Helv.*, **29**, No. 1, 9–26 (1955).
3. "Faisceaux algébriques cohérents," *Ann. Math.*, **61**, No. 2, 197–278 (1955).
4. "Géométrie algébrique et géométrie analytique," *Ann. Inst. Fourier*, **6**, 1–42 (1955-1956).
5. *Algèbre Locale. Multiplicités*, Springer-Verlag, Berlin–New York (1975).

I. R. Shafarevich

1. *Foundations of Algebraic Geometry* [in Russian], Nauka, Moscow (1972).

A. Silva

1. "Rungescher Satz and a condition for Steinness for the limit of an increasing sequence of Stein spaces," *Ann. Inst. Fourier*, **28**, No. 2, 187–200 (1978).

Y.-T. Siu

1. "A proof of Cartan's theorems A and B," *Tohoku Math. J.*, **20**, No. 2, 207–213 (1968).
2. "Analytic sheaf cohomology groups of dimension n of n-dimensional complex spaces," *Trans. Am. Math. Soc.*, **143**, Sept., 77–94 (1969).

E. G. Sklyarenko

1. "Homology theory and the exactness axiom," *Usp. Mat. Nauk*, **24**, No. 5, 87–140 (1969).

O. Stanasila

1. "On the dualizing complex," Romanian–Finnish Seminar on Complex Analysis, *Lect. Notes Math.*, **743**, 475–482 (1979).

K. Stein

1. "Analytische Funktionen mehrerer komplexer Veränderlichen zu vorgegebenen Periodizitätsmoduln und das zweite Cousinsche Problem," *Math. Ann.*, **123**, No. 2, 201–222 (1951).

K. Suominen

1. "Duality for coherent sheaves on analytic manifolds," *Ann. Acad. Sci. Fenn.*, No. 424, 1–19 (1968).
2. "Localization of sheaves and Cousin complexes," *Acta Math.*, **131**, No. 1–2, 27–41 (1973).

J.-L. Verdier

1. "Topologie sur les espaces de cohomologie d'un complexe de faisceaux analytiques à cohomologie cohérente," *Bull. Soc. Math. Fr.*, **99**, No. 4, 337–343 (1971).
2. "Catégories dérivées," Séminaire de Géométrie Algébrique, *Lect. Notes Math.*, **569**, 262–311 (1977).

V. Villani

1. "Un teorema di passaggio al limite per la coomologia degli spazi complessi," *Atti Accad. Naz. Lincei*, **43**, No. 3–4, 168–170 (1967).

V. S. Vladimirov

1. *Methods of the Theory of Functions of Several Complex Variables* [in Russian], Nauka, Moscow (1964).

A. Weil

1. "Sur les théorèmes de de Rham," *Commun. Math. Helv.*, **26**, No. 2, 119–145 (1952).

H. Whitney and F. Bruhat

1. "Quelques propriétès fondamentales des ensembles analytiques-réels," *Comment. Math. Helv.*, **33**, No. 2, 132–160 (1959).